D1826351

Striated muscle is the most common muscle type in the vertebrate body. This book describes in molecular terms the components and intracellular events responsible for the contraction and relaxation of striated muscle. The topic is introduced with a discussion of motile systems occurring throughout the biological world and their relation to the highly specialised contractile system of muscle. Professor Perry then goes on to discuss the mechanochemical process and the regulatory roles of calcium, I filament proteins and phosphorylation. The book ends with an examination of the role of dystrophin and its implications in Duchenne muscular dystrophy, the most common form of muscle disease.

   Molecular Mechanisms in Striated Muscle will provide an important source of information and current theory for researchers and postgraduate students in muscle physiology, biochemistry and medicine.

Molecular Mechanisms in Striated Muscle

Lezioni Lincee
Sponsored by *Foundazione IBM Italia*
Editor: *Luigi A. Radicati di Brozolo, Scuola Normale Superiore, Pisa*

The Lezioni Lincee arises from lectures given under the auspices of the Accademia Nazionale dei Lincei and is sponsored by *Foundazione IBM Italia*.

The lectures, given by international authorities, will range on scientific topics from mathematics and physics through to biology and economics. The books are intended for a broad audience of graduate students and faculty members, and are meant to provide a '*mise au point*' for the subjects with which they deal.

The symbol of the Accademia, the lynx, is noted for its sharpsightedness; the volumes in the series will be penetrating studies of scientific topics of contemporary interest.

Already published

Chaotic Evolution and Strange Attractors: D. Ruelle
Introduction to Polymer Dynamics: P. de Gennes
The Geometry and Physics of Knots: M. Atiyah
Attractors for Semigroups and Evolution Equations:
    O. Ladyzhenskaya
Asymptotic Behaviour of Solutions of Evolutionary Equations:
    M.I. Vishik
Half a Century of Free Radical Chemistry: D.H.R. Barton in
    collaboration with S.I. Parekh
Bound Carbohydrate in Nature: L. Warren
Neural Activity and the Growth of the Brain: D. Purves
Perspectives in Astrophysical Cosmology: M. Rees

Molecular Mechanisms in Striated Muscle

S. V. PERRY

Emeritus Professor of Biochemistry, Department of Physiology,
University of Birmingham

CAMBRIDGE
UNIVERSITY PRESS

Published by the Press Syndicate of the University of Cambridge
The Pitt Building, Trumpington Street, Cambridge CB2 1RP
40 West 20th Street, New York, NY 10011-4211, USA
10 Stamford Road, Oakleigh, Melbourne 3166, Australia

© S. V. Perry 1996

First published 1996

Printed in Great Britain at the University Press, Cambridge

A catalogue record for this book is available from the British Library

Library of Congress cataloguing in publication data

Perry, S. V.
Molecular mechanism in striated muscle / S. V. Perry.
    p.   cm. – (Lezioni lincee)
Includes bibliographical references and index.
ISBN 0 521 57001 8 (hc).   ISBN 0 521 57916 3 (pb)
1. Muscles – Molecular aspects.   2. Muscle contraction.   I. Title.
    II. Series.
QP321.P456   1996
596'.01852 – dc20   95-50334 CIP

ISBN 0 521 57001 8 hardback
ISBN 0 521 57916 3 paperback

V N

Contents

Preface

The origins of this book lie in a series of lectures given in Rome, Ancona, Pardua and Pisa in the summer of 1990 under the auspices of the Accademia Nationale dei Lincei. The material presented in these lectures has been extended and updated but the book does not claim to be an all-embracing account of the subject. It covers aspects of the biochemistry of muscle in which I have had a special interest throughout my academic career. I am particularly grateful to Barry Levine, Val Patchell and Phil Quirk who read sections of the manuscript and made comments that led to its improvement.

1

General aspects of motile systems

Directed movement as a result of force development is a characteristic of all living organisms. Random movement of molecules and particles is a consequence of the thermal energy of the system and is responsible for the Brownian movement of small organelles. Such processes are responsible for the diffusion of metabolites and ions in the cell and their equilibration within regions bounded by membranes with differential permeabilities. Normal activity requires controlled movement at the different levels of biological organisation ranging from defined regions within the cell to whole tissue. This need has led to the evolution of a number of systems designed to bring about force development and directed movement appropriate to these different demands at the expense of metabolic energy.

Actin-based motor protein systems

The major contractile tissue responsible for movement in the animal kingdom is muscle. This tissue is composed of highly specialised cells in which the components responsible for contraction and its regulation and for producing adenosine triphosphate (ATP), the substrate for contraction, represent up to 90% or more of the total mass. The contractile properties of the muscle tissues have adapted to the different functions they have to perform. These include the movement of limbs, the circulation of blood, control of the vascular system and many other functions. Accommodation to these different requirements has resulted in further specialisation so that we have, ranged in order of decreasing specialisation, three types of muscle tissue: skeletal, cardiac and smooth. Although these muscle cell types exhibit obvious structural differences they all possess a very similar contractile system. In each case contraction occurs by relative movement between the two protein filaments that make up the actomyosin system. The thicker of the two filaments is built up by linear aggregation of molecules of the protein motor, myosin, which has a rod-like structure with two globular heads (Fig. 1.1). Parallel to the thick filament lies the thin filament composed of two linear arrays of actin monomers arranged in a double helix. The interaction between the myosin globular heads and the monomer units of actin which is associated with the hydrolysis of ATP leads to lateral movement and tension development (see Chapter 2).

From the early studies it was clear that the molecular structure and protein composition of the contractile systems of all

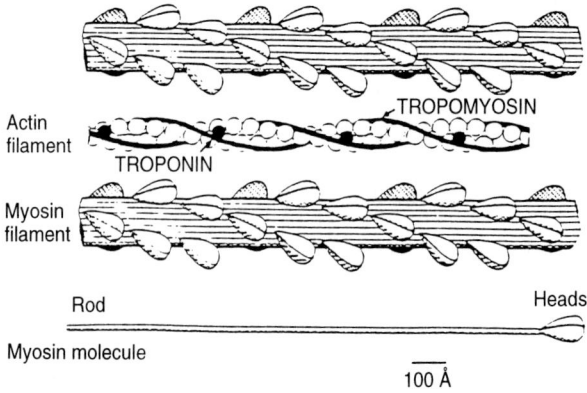

Actin filament — TROPOMYOSIN

TROPONIN

Myosin filament

Rod   Heads

Myosin molecule

100 Å

Fig. 1.1 Scheme illustrating the structures of the myosin and actin filaments in vertebrate skeletal muscle. In the thicker filaments the myosin II heads are arranged on a three-stranded helical array. The troponin complex involved in the regulation of striated muscle is not present in smooth muscles and other actomyosin contractile systems that have different regulatory mechanisms. Reproduced from Squire (1983).

types of muscle were remarkably similar. Of the two components the actin is the most strongly conserved. The relatively small number of isoforms that occur in the various muscle types and non-muscle tissues in a given species differ conservatively in very few residues. Furthermore, actins from vertebrates, yeast and amoeba are remarkably similar in amino acid sequence. The muscle myosins are always double-headed and are indistinguishable when observed by electron microscopy. Nevertheless despite their similarity in overall structure and biochemical properties, the myosins in a given species occur in a much larger number of isoforms than is the case with actin, as many as 20–30 being reported in the muscles of some vertebrates. The slight differences in sequences of the myosin isoforms are responsible for fine tuning of the biochemical properties of the motor, for example the ATPase activity, to accommodate the particular physiological function of the muscle type. Experience with muscle contractile systems suggested that an essential component of the actomyosin contractile system was a double-headed myosin molecule. Indeed, the muscle two filament array was taken as the prototype contractile system.

The discovery by Pollard & Korn (1973) that *Acanthamoeba* contained in addition to the normal type, a myosin of lower molecular weight with only one head widened views on the myosin family of proteins and their role in the contractile process. By convention it has been customary to identify the single-headed motor protein responsible for a variety of motile functions, particularly in non-muscle cells, as myosin I and the two-headed form characteristic of muscle tissue and some other contractile functions in non-muscle cells such as cytokinesis and receptor capping in *Dictyostelium* as myosin II. The existence of myosin I as a protein with contractile function and the demonstration that single heads of myosin II, subfragment I, could move along actin filaments in model motile systems clearly indicates that two heads on the myosin molecule are not required for contraction to occur.

With the advances in molecular biology in recent years there has been an explosion of knowledge about the myosin motor superfamily in eucaryotic cells. This has resulted in the identification of additional myosin classes that are distinguished by the numerals III to XI, allocated in order of discovery after myosins I and II (Hammer, 1994; Mooseker & Cheney, 1995). The amino acid sequences of these newly discovered classes differ from one another as much as that of myosin I differs from myosin II. Myosins other than the conventional double-headed protein associated with muscle, myosin II, are often referred to collectively as unconventional myosins. Not all of the unconventional myosins have been isolated and in these cases the evidence for their existence is from identification of the gene, cloning the cDNA and determining its sequence. The amounts of these proteins present in the cell are usually small compared with those of myosin II in contractile tissues even though they are widely distributed in animals and in plants. The genes have been identified by consensus sequences associated with the myosin head. Despite possessing domains characteristic of myosin, the different myosin types also exhibit considerable variations in sequence which no doubt are of significance for the particular function of each myosin motor (Table 1.1). A feature of the unconventional myosins is that they possess very much shorter tail regions and as a consequence they do not usually form filaments like the conventional myosins. In many cases the tail regions bind, or are predicted to bind, to membranes. Thus one function of these molecules is to provide a contractile link between membranes and the actin-rich cortical structures that are involved in cell shape and locomotion. Their properties and localisation in the cell are such as to suggest that they are also involved in the movement of membranes, particles and vesicles about the cell, cytoplasmic streaming, psuedopodia extension, budding, formation and the maintenance of microvilli and many other cellular functions that depend upon contractile systems. Studies with *Dictyostelium* have thrown some light on the different roles of the myosins in this organism. By inhibiting expression of the gene it has been demonstrated that during development myosin II is implicated in cytokinesis and change in cell shape, but surprisingly not in migration along surfaces (for a review of the myosin motor, see Spudich, 1994).

One particularly interesting feature of all myosin types is the presence in the neck region of the head (Chapter 2) of a consensus sequence IQXXXRGXXXR, the so-called IQ motif. The frequency of this sequence in the unconventional myosins varies from one, for example in the high molecular weight unconventional myosin of *Acanthamoeba*, to six in the P190 unconventional myosin of chicken brain. In the conventional myosins there is one IQ motif together with a less well conserved version of it in the α-helical region of the neck, where the light chains are bound. The IQ motif shares sequence similarity with the calmodulin-binding domains of non-myosin proteins and it has been suggested (Cheney & Mooseker, 1992) that they represent binding sites on myosin for members of the calmodulin/E-F hand superfamily of proteins. This α-helical region where the light chains and calmodulin bind is considered to be important for contractile function and its regulation. In the most highly specialised myosin II of striated muscle the light chains have preserved their role in the transduction systems but their role in regulation is much attenuated for this is largely initiated by systems associated with the actin filament (Chapters 3 and 4). With the more primitive form of myosin II present in smooth muscle and the unconventional myosins generally the low molecular weight proteins associated with the head have preserved their regulatory function. In these cases the myosin motor is regulated by the direct influence of calcium on the light chain or calmodulin or by its indirect influence in the activation of kinases that phosphorylate the light chains. In this respect it is of interest that in the NinaC unconventional myosin present in photoreceptor cells of *Drosophila* a kinase-like domain can be identified as part of the head sequence (Table 1.1).

Table 1.1. *Schematic diagram of the different classes of the myosin motor superfamily*

Class	Properties	Examples
Conventional myosins: Myosins II	Form dimers and bipolar filaments; responsible for muscle contraction, required for cytokinesis and receptor capping in *Dictyostelium*	Skeletal muscle MII, smooth muscle MII, non-muscle MII, *Dictyostelium* MII (~200 kDa)
Unconventional myosins: *Dilute*/P190/*MYO2*	Form dimers but not filaments; the *myo2* mutant is defective in budding, has disorganised actin and accumulates vesicles	Mouse *Dilute*, chicken brain, p190, yeast *MYO2* (180–215 kDa)
Ameboid myosins I with tails that bind to membranes and actin	Form monomers that can cross-link actin; localised to small vesicles, phagocytic cups, the contractile vacuole and the leading edges of moving cells	*Acanthamoeba* MIA, MIB, MIC; *Dictyostelium* MIB, MID (125–140 kDa)

Coiled-coil α-helix

Globular domains

Coiled-coil α-helix

Membrane binding

ATP-independent actin binding

Membrane binding	Ameboid myosins I with tails predicted to bind to membranes	Unknown	*Dictyostelium* MIA, MIE (~113 kDa)
Membrane binding	Vertebrate myosin I that bind to membranes and have three to four calmodulin light chains	Form monomers; link actin bundles to the plasma membrane in the microvillus	Brush border MI, brain/adrenal MI (~120 kDa)
?	*Acanthamoeba* high molecular weight unconventional myosin	Unknown	*Acanthamoeba* high molecular weight MI (177 kDa)
Kinase-like domain	NinaC	Present in *Drosophila* photoreceptor cells; the 174 kDa form is localised to the rhabdomere where it may function in phototransduction	NinaC splice forms (174 and 132 kDa)

The myosin head domains are represented by large open regions and the myosin light chains by small filled ovals. M, myosin.

Source: from Cheney & Mooseker (1992).

The greatest variability in the sequences of the myosin classes is found in the tail regions. This is a consequence of the differing roles that the myosin motor plays in general cell activity. Conventional myosins are characterised by the long double α-helical tail which facilitates polymerisation into strong filaments to develop and maintain the high tensions required for muscle function. In many other cell functions the myosin operates as a single molecule with a much smaller tail for attachment to other cell structures and proteins. Despite the variation in tail sequences between the different myosin classes, within a myosin class the tail sequences are related which suggest that the head and tail sequences of the myosins are coupled during evolution.

The myosin superfamily offers many opportunities for phylogenetic analysis which has confirmed the classification based on the sequences of the head regions. The analyses are consistent with the existence of at least 11 classes of myosin to which reference has already been made. Study of the myosin superfamily suggests that there existed an ancestral form from which all myosins evolved. Current knowledge does not permit definite conclusions to be made as to the precise nature of this myosin but some generalisations can be made. It was likely to be single-headed with calmodulin-binding sites in its head and present in non-muscle tissue from which muscle tissue evolved. The phylogenetic data suggest that striated muscle is older than smooth muscle and that these types developed independently from non-muscle tissue (Goodson & Spudich, 1993). Support for this view comes from the fact that the light chains present in the myosin II are derived from calmodulin or its immediate precursor (Collins, 1991). This suggests that when the myosin evolved into the double-headed form associated with muscle tissue the proteins associated with the IQ motifs in the head took on a different function in the regulatory process for they are no longer able to bind calcium with the high affinity exhibited by calmodulin.

Tubulin-based motor protein systems

As outlined above the myosin family of motor proteins plays an important part in a number of general cell activities in addition to their highly specialised role in muscle tissue. Although the nature of the contractile activity in which the myosin motor is involved may vary, in all cases the filament along which it moves is the linear actin polymer. It is now clear that a number of other motor proteins are employed by eucaryotes to carry out a variety of motile functions such as the movement of chromosomes, vesicles, other intracellular components and the beating of cilia and flagella, etc. These systems bear some resemblance to the actomyosin system in that the motor protein is an enzyme hydrolysing ATP to provide the energy to move unidirectionally along a protein filament. In these cases the linear track along which the motor moves is the microtubule, a widely distributed intracellular structure. The first motor protein associated with tubulin to be identified was named dynein which was shown to be located in 'arm-like' projections along microtubules present in the axonemes of cilia of *Tetrahymena* and the flagella of sperm from various organisms. Dynein and tubulin form the bulk of these motile structures and it was originally considered that the dynein–tubulin system was unique to axonemes. It is now known that both proteins are widely distributed in cells although in smaller amounts than are present in the highly specialised axoneme (for a review, see Walker & Sheetz, 1993).

Microtubules are hollow cylindrical polymers about 25 nm in diameter that are found in virtually all eucaryotic cells (Fig. 1.2). They are built up from tubulin, a globular dimeric protein of molecular mass 110 kDa, that is composed of two non-identical subunits, α and β. Tubulin is strongly conserved like actin but there are many more distinct tubulin genes. Like actin filaments microtubules exhibit a so-called 'plus' end where polymerisation occurs rapidly and a 'minus' end where polymerisation is slow. In most cells the 'plus' end corresponds to the distal end, that is the further removed from the cell centre, and the 'minus' to the proximal end. The fact that protein motors can move in either direction along the micro-tubules, despite their fixed polarity, is an important difference from the myosin motor which invariably moves to the barbed or 'plus' end of the actin filament (Chapter 2). The great similarity between the properties of actin and tubulin prompted the suggestion that they evolved from the same gene but comparison of their amino acid sequences indicates that this is not the case.

Dynein

Dyneins are a family of motor proteins with ATPase activity stimulated by microtubules in a manner that is analogous to the stimulation of the MgATPase of myosin by actin. Movement of the dynein motor along the tubulin filament towards the 'minus' end accompanies the hydrolysis of ATP. They are high molecu-

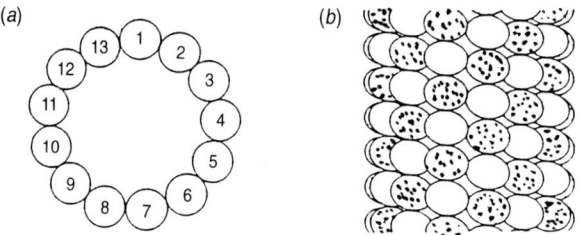

Fig. 1.2. Diagrammatic representation of the arrangement of tubulin subunits in the cytoplasmic microtubule. (*a*) Cross-section of a microtubule showing arrangement of 13 protofilaments. (*b*) View of the surface lattice of the microtubule showing the arrangement of the α and β subunits, represented by stippled and clear ovoids, respectively. The subunits in the microtubule are arranged alternately along the protofilament and a three-start helix. Reproduced from Snyder & McIntosh (1976).

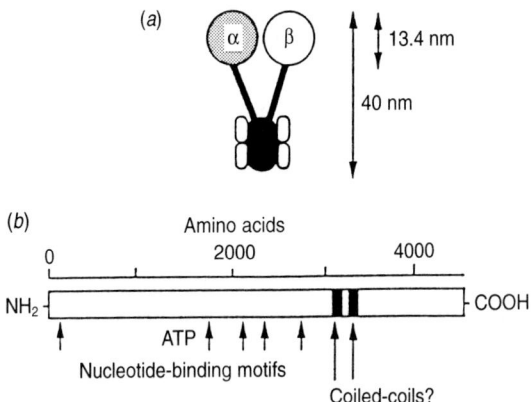

Fig. 1.3. (*a*) Schematic representation of flagellar outer arm dynein. (*b*) Domains of the ~500 kDa heavy chain of dynein. Reproduced from Vale (1992).

lar mass proteins of about 500 kDa that contain several consensus sites for nucleotide binding, only one of which appears to be involved in the hydrolysis of ATP. Their motile properties are very similar but the structure of the dyneins varies according to the source. Different forms are present in the outer and inner arms of ciliary and flagellar axonemes for example (for a review, see Vale, 1992). The dyneins may contain one to three heavy chains and a corresponding number of heads when observed by electron microscopy. In addition there are several types of associated polypeptides, including light chains. Some of these polypeptides are located near the base of the molecule as is the case with the flagellar outer arm dynein, illustrated in Fig. 1.3.

Kinesin

Kinesin was discovered as a microtubule-associated motor protein responsible for vesicle transport in neuronal cells. It shares many properties with dynein in that it possesses ATPase activity stimulated by microtubules in the transduction process during which it usually moves to the 'plus' end of tubulin. Its gene was cloned soon after its purification enabling the identification of the motor domain sequence of 350–400 amino acids containing the ATPase and tubulin binding sites. Genetic screening for the motor domain sequence has revealed the existence of a large family of kinesins and kinesin-related proteins that are widely distributed in cells.

In its conventional form kinesin is an elongated molecule with two globular domains attached to a long stalk with a fan-like tail domain (Fig. 1.4). Like myosin II the molecule is built up of two heavy chains that associate to form an α-helical coiled-coil domain in the stalk region. The heavy chains separate in the N-terminal region to form the two globular regions which represent the motor domains where the ATP and microtubule binding sites are located. Light chains are bound at the small carboxy-terminal tail but as yet these have not been well characterised and are somewhat variable in number. The kinesin molecule is significantly shorter because the heavy chains contain about half the number of residues present in the

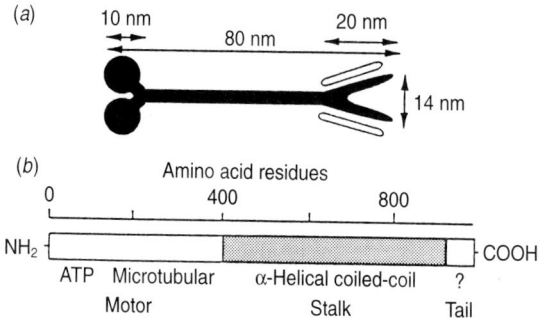

Fig. 1.4. (*a*) Schematic representation of conventional kinesin which is composed of two identical heavy chains of ~110 kDa with light chains associated with the fan shaped tail region. (*b*) Domains of the kinesin heavy chains. Reproduced from Vale (1992).

corresponding structures in myosin II. It is probable that the heavy and light chains in the tail region are involved in linking kinesin to the specific intracellular structures that it translocates along microtubules. A single immobilised kinesin molecule can transport microtubules or beads for several microns indicating that it is a highly progressive motor that remains bound to microtubules for the majority of its ATPase cycle (Vale, 1992). The kinesins are classified from the consensus sequence of the motor domain but there is no homology between the different members of the kinesin superfamily in the amino acid sequences of the non-motor domains. These differences in structure presumably reflect, as is the case with the myosin motors, the variety of roles of the kinesins and their ability to interact with different structures in the cytoplasm. In addition to their role in axonal transport of vesicles the kinesins have been implicated in chromosome movement during mitosis and meiosis, separation of the spindle pole-bodies during mitosis and nuclear migration. No doubt they are involved in other functions requiring translocation of intracellular components.

Dynamin

Dynamin, which was first identified in calf brain, possesses GTPase that is activated by microtubules and belongs to a superfamily possessing a GTP-binding domain in the N-terminus (Shpetner & Vallee, 1992). It differs from dynein and kinesin in that it consists of a single polypeptide chain of 100 kDa. The 30 kDa N-terminus forms a globular domain which contains the GTP-binding consensus sequence and the microtubule-binding domain is located in a C-terminal region (Fig. 1.5). The precise role of dynamin has yet to be determined. It possesses homology with proteins involved in vacuolar protein sorting and spindle-body separation. The similarity of the effects of microtubules on its ATPase activity with those obtained with dynein and kinesin suggest that it has a motor

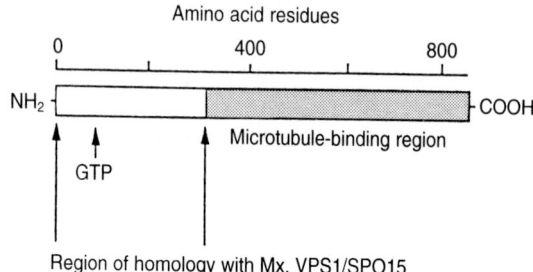

Fig. 1.5. Schematic representation of the domains of the 100 kDa polypeptide of dynamin. Although the evidence is suggestive further work is necessary to demonstrate conclusively that dynamin is a force-generating protein. Reproduced from Vale (1992).

function. The defects in endocytosis and recycling of synaptic vesicles that arise in mutation studies indicate a role in vesicle budding (Hinshaw & Schmid, 1995).

General features of eucaryotic motor proteins

All the proteins so far discovered with clearly defined motor function possess an N-terminal globular domain containing sites that bind the substrate for transduction, usually ATP. Associated with the motor protein is the protein structure that forms the track along which movement proceeds. The F-actin filament and microtubule both exhibit polarity and the direction of movement along them is specific to the motor protein. The myosin II present in muscle invariably moves to the 'plus' or barbed end, an essential requirement for myofibrillar contraction. It is probable that all myosins move in the same direction with respect to the actin sequence in view of the fact that they are identified by the consensus sequence of the myosin motor. Detailed investigation of the unconventional myosins, however, could well lead to some surprises in view of the experience with the tubulin motor systems.

There does not appear to be a universal consensus sequence that is unique to the functional domains of all the motor proteins. Indeed the functional domains of motor proteins would be expected to be in part determined by the actin or tubulin track along which they move. It is perhaps more surprising that the motor domain for dynein should differ enough from that of kinesin, both using ATP and binding to microtubules, to enable the families to be distinguished. This may relate to the fact that usually locomotion with dynein is to the 'minus' end whereas that of kinesin is to the 'plus' end. Occasional exceptions to these rules have been reported. For example dynein-like motor activity from *Reticulomyxa in vitro* has been observed to take place in both directions unless the dynein is phosphorylated when it moves to the usual 'minus' end (Euteneuer *et al.*, 1988). Also the claret segregation protein of *Drosophila* with a similar molecular weight and motor domain sequence to the kinesins moves towards the 'minus' ends of the microtubule. Thus it would appear that despite the fixed polarity of the protein filaments it is the structure of the domains on the motor proteins that finally decides the direction in which locomotion proceeds.

One of the puzzling aspects of the motor proteins is their multiplicity, probably as many as 100 different forms in a cell (Spudich, 1994). A possible explanation is that each motor protein has evolved so that its contractile properties are specific for the role that it has to perform. As has been pointed out earlier actin and tubulin are very similar throughout the animal kingdom and each protein is virtually identical in properties in all the locomotory systems in which it is involved. It makes biological sense to modulate the properties of the system by slight changes in the structure of the motor protein responsible for the transduction process. This feature has long been apparent in the widely studied myosins of muscle. Each muscle has a myosin II isoform composition that is specific for the cell type and its stage of development. For example myosin from a fast muscle has a higher V_{max} for the ATPase than that from a slow skeletal muscle. Also changing the speed of a muscle by changing the innervation induces a change in gene expression so that myosin isoforms appropriate for the altered speed of the muscle are produced. In these circumstances the actin does not change because the fast and slow skeletal muscles have the same isoform. Nevertheless, a tight association of motor protein with function does not appear to apply with all the microtubule-based motors. There are examples of dynein and kinesin-based motors performing several functions and yeast mutants lacking certain kinesin-like proteins appear to be viable and show little change in phenotype.

Despite the differences that exist between the three main classes of motor proteins, which are largely of a functional nature, the stimulation of ATPase activity by the track protein and the coupling of ATP hydrolysis with the lateral movement between the two protein systems strongly suggest a very similar mechanochemical process. The kinetics of ATP hydrolysis by myosin II in the muscle actomyosin system and its relationship to force production has been studied in detail (Chapter 2). The conclusion from these studies is that force production is associated with phosphate release following ATP hydrolysis and that dissociation of actomyosin is tightly coupled to the binding of ATP by myosin. The dynein–microtubule ATPase system exhibits many similarities to actomyosin implying that these motor proteins have a common mechanism. On the other hand the kinesin system does appear to possess features that are different. Recent studies by Romberg & Vale (1993) on the translocation of microtubules along kinesin attached to a microscope slide indicate that, in contrast to the myosin and dynein motor systems, kinesin with bound ADP dissociates from microtubules during translocation whereas kinesin with ATP bound remains tightly bound to the microtubules. Also, Berliner *et al.* (1995) have reported that kinesin requires two active heads to track a microtubule protofilament and conclude that the two heads act in a coordinated 'hand-over-hand' mechanism. If this is the case it suggests that the mechanochemical process differs from the myosin system where current evidence suggests the two heads act independently. Thus despite the common features of all the protein motor systems, which are rather striking, it would appear that the mechanism of the kinesin motor is not identical to that of myosin and dynein (Taylor, 1993; Schnapp, 1995). Recent three-dimensional structural studies on microtubules and tubulin sheets decorated with kinesin are beginning to define the conformational changes in tubulin and kinesin that accompany motor function (Hoenger *et al.*, 1995). Of particular interest in relation to the actomyosin system is the demonstration of a nucleotide-dependent change in the kinesin motor domain bound to tubulin (Hirose *et al.*, 1995).

For a translocation system to work effectively it must be regulated. With the exception of myosin II (see Chapters 3, 4 and 5) relatively little is known of the mechanisms involved. Calcium is the trigger that initiates contraction in the highly specialised myosin II systems present in muscle. Nevertheless the events that subsequently lead to activation of the ATPase are not the same in different muscle types. The persistence of calmodulin-binding sites associated with the motor domains of the families of unconventional myosins suggest that these proteins also are regulated by calcium. The mechanisms are uncertain but may involve the dissociation of calmodulin from the binding sites that occur in the presence of calcium. If this is the case it is a different mechanism from that applying to myosin II where the light chains remain attached to the head even if calcium is bound, as is the case with molluscan adductor myosin (see Chapter 5). Very little is known about the regulation of the microtubule motors but phosphorylation may be involved as it is known to occur on the heavy and light chains of kinesin.

Motility in procaryotes

Procaryotes, presumably because of their small size and limited structural organisation, have not developed a range of intra-cellular motile systems comparable to those of the eucaryotes. The latter usually involve movement of the protein motor unidirectionally along a protein track with the result that in effect the actin filament or microtubule slides past the motor and its attached structure. Like some eucaryotes, procaryotes have adapted the flagella for cellular motility. Whereas eu-caryotes use a sliding movement between the microtubules powered by the dynein system to produce the undulating action of flagella that propels the cell along, the procaryotic flagellum is an entirely different structure powered by a different kind of motor system which produces rotation of a protein filament (for a review, see Armitage, 1992). Rotation of actin filaments has been reported in an *in-vitro* motile system (Nishizaka *et al.*, 1993) but its physiological significance is uncertain and it is a rare event as is the case with microtubular systems. There is a report of the kinesin-like claret segregation protein rotating round the microtubule as it moves along, thus having the potential for producing torque. The ability to produce torque in microtubules has also been reported for a 14S ciliary dynein from *Tetrahymena* (Endow, 1991).

The bacterial flagellum is a relatively inert system compared with that of the eucaryote. The flagellum is composed of a single protein, flagellin, polymerised into a flexible tube-like structure, stiffer than an F-actin filament, which when rotated develops thrust in a propeller-like manner. Rotation takes place by the action of a unique structure just inside the cell known as the basal body (Fig. 1.6). The flagellum is connected to the flagellar rotory motor in the basal body by the hook, a polymerised protein structure. The flagellar rotatory motor is driven by proton motive force in many procaryotes but in some cases by other ion flow, for example in *Vibrio alginolyticus* by the movement of sodium from the outside to the inside of the cell. With an overall diameter of about 25–50 nm the proton motor is the smallest rotatory mechanical device known. Its size should be compared with the fundamental contractile step in the cross-bridge cycle of muscle. The movement of a single myosin head associated with the hydrolysis of one molecule of ATP is estimated to be about 10–15 nm. It seems probable that the linear movement associated with a single cycle in the dynein protein motor system is similar. In the proton motor it is probable that movement occurs as the result of conformational changes causing two proteins, one associated with the stator and the other with the rotor, to change position relative to each other, as is the case with eucaryotic motor proteins. In view of the probable similarity of mechanism at the molecular level it is of interest to compare the sliding velocities between the two components in the two types of motile system. It can be

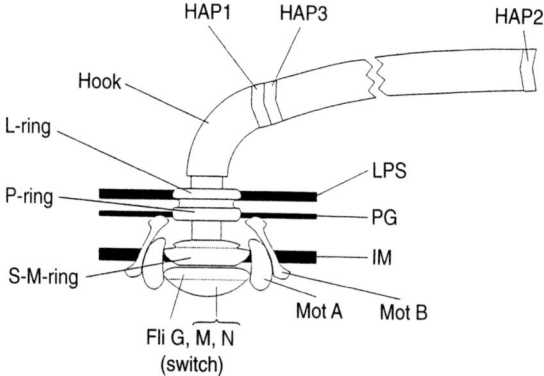

Fig. 1.6. Schematic representation based on electron micrographs of the proton motor and its relation to the bacterial cell membrane and the flagellum. The diameter of the filament which is connected to the motor by the hook is about 17 nm. The whole motor complex is about 50 nm. HAP, hook associated proteins; IM, inner membrane; PG, peptidoglycan; LPS, lipoylsaccharide. Reproduced from Armitage (1992).

calculated that myosin moves along the actin filament at about $10\,\mu m\ s^{-1}$ in skeletal muscle and around $100\,\mu m\ s^{-1}$ in the cytoplasmic streaming of plant cells such as *Chara* and *Nitella*. The rotation rates of the procaryotic flagellar motors are high, ranging from 170 rev s^{-1} in *Salmonella typhimurium* to up to 1700 rev s^{-1} in *Vibrio alginolyticus*. The latter speed would correspond to a sliding velocity between the rotor and stator of $160\,\mu m\ s^{-1}$ (Magari *et al.*, 1994). Thus the sliding velocities observed with the motor proteins of procaryotes lie close to or within the range obtained with the actomyosin systems. The flagellar motor does, however, appear to differ from the eucaroytic motors in that it can be readily reversed, for example by changing from clockwise to anticlockwise rotation of the flagellar motor the procaryote cell can change its direction of movement.

Thus it can be concluded that molecular mechanisms for force production in all living systems exhibit a number of similarities. This suggests that the basic mechanism for force development is common to all systems, adapted to accommodate the special requirements of the role it has to play. Despite the basic similarities that must underlie the mechanochemical process, the form of energy used by it in eucaryotes and procaryotes is different. Because of their smaller size and simpler cell organisation and function many procaryotes do not require complex intracellular translocation systems and the specialised motor proteins required for their function. Their major requirement is for a mechanism that will enable the whole cell to move in a controlled manner. To energise the bacterial flagellum, protein gradients, universally used in biological systems for the production of utilisable energy, have been adapted to drive the proton motor directly. Proton gradients also provide much of the energy required for the motile systems of eucaryotes but not directly. The multiplicity of motile functions widely dispersed throughout the cell necessitates the eucaryotes having an energy source freely available in all regions of the cell, and not only at a membrane structure, with the appropriate properties to harness the energy of proton or ion flow. Thus ATP (and possibly in some cases GTP) which is produced centrally by proton flow in the mitochondria is used universally to drive the eucaryotic motor proteins. In this respect it is probably significant that there are many lines of evidence suggesting that mitochondria evolved as a consequence of bacterial capture by the eucaryotic cell at an early stage in its development.

Muscle tissue is by far the most widely studied motile tissue and a wealth of detail regarding its nature and function exists. In the chapters that follow the molecular basis of the contractile process and its regulation in the different muscle types will be discussed in the conviction that an understanding of this highly specialised function is essential to an understanding of the contractile process in its various forms.

2

Mechanochemistry of contraction

The most highly specialised of the actomyosin systems is that present in striated muscle where it is organised to develop high tensions and to shorten rapidly. For this purpose it is built into a fixed structure, the myofibril (Fig. 2.1), which makes up 50–60% of the total muscle cell protein, the highest amounts being present in fast twitch muscle. In addition to actin and myosin, which together make up about 75% of the total mass, the myofibrils contain a number of other proteins (Table 2.1). As model contractile systems can be made from purified actin and myosin these additional proteins are not essential for the contractile process *per se* but either are involved in its regulation or have a structural role in the myofibril and for its relation to the cytoskeleton. This chapter is concerned with the nature of

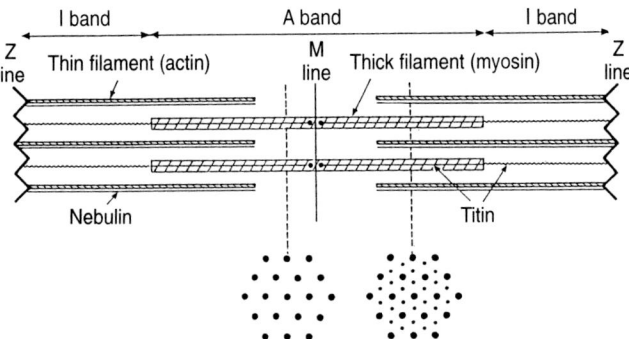

Fig. 2.1. Diagrammatic representation of the myofibril of vertebrate striated muscle illustrating the arrangement of the filaments when seen in cross-section.

Table 2.1. *Major protein components (present in 2% or greater amounts) of the myofibril of skeletal muscle*

Protein	Localisation	Molecular weight ($\times 10^3$)	Amount (% wt)
Contractile			
Myosin	A band	520	53
Actin	I, A bands	42 (monomer)	22
Regulatory			
Tropomyosin	I, A bands	66 (dimer)	5
Troponin C	I, A bands	18 ⎫	
Troponin I	I, A bands	21 ⎬	5
Troponin T	I, A bands	31 ⎭	
Structural–cytoskeletal			
Titin (connectin)	I band	2100	10
Nebulin	I band	800	5
α-actinin	Z line	190 (dimer)	2
C protein	A band	135	2
M protein	M line	165	2

Values for the contractile and regulatory proteins have been obtained by calculating the number of molecules present from the known dimensions of the filaments and assuming that actin and myosin represent 75% of the total protein. The value for the total amount of the structural-cytoskeletal proteins (Otsuki *et al.*, 1986) is in the view of the author likely to be high. A number of other proteins each representing less than 1% of the myofibril are present (Otsuki *et al.*, 1986). Most of the values relate to the proteins of rabbit fast skeletal muscle.

the interaction between actin and myosin that is fundamental to the contractile process.

Actin

Actin is a very widely distributed protein of molecular mass 42 kDa found in practically all eucaryotic cells. The monomeric form known as G-actin polymerises in the presence of divalent cations and at higher ionic strengths to F-actin, the form involved in the contractile process. F-actin exists as a filament composed of two strands produced by linear aggregation of G-actin subunits that are organised into a double helix with a pitch of 77 nm and a diameter of 8 nm (Fig. 2.2), because of the double-stranded structure, the filament has an apparent repeat of 38.5 nm, half the pitch. It is of some interest that a molecule of

(*a*)

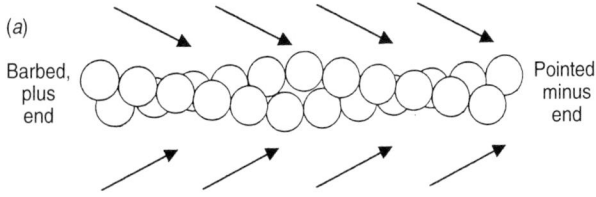

Barbed, plus end

Pointed, minus end

(*b*)

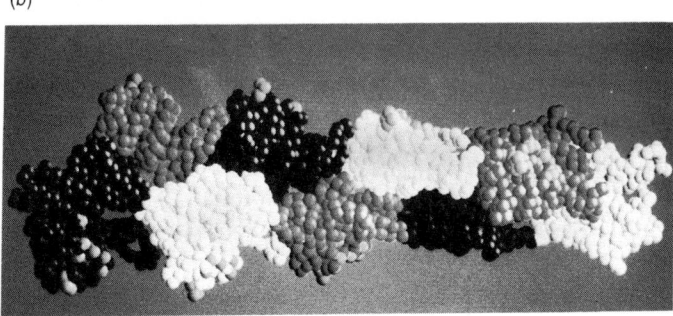

Fig. 2.2. (*a*) Diagrammatic representation of the structure of the F-actin filament. Circles represent actin monomers of 42 kDa. Arrows indicate the direction in which the myosin molecules orient themselves when the heads interact with the actin monomers, that is, when the filament is 'decorated'. (*b*) Model of actin filament obtained by fitting the structure of the actin monomer to the observed X-ray diffraction pattern of oriented F-actin. Reproduced from Holmes *et al.* (1990).

ATP is associated with each actin monomer. This is known as the bound nucleotide and is dephosphorylated to ADP when G-actin is polymerised to F-actin. There is no evidence that the bound nucleotide is involved in any metabolic role; indeed, the ADP bound to F-actin is not readily available to enzymic systems for which it normally is a substrate. In F-actin the monomers are arranged with a precise polarity that is similar in both filaments and is of importance in establishing the manner in which myosin interacts with the filament. Thus when F-actin filaments are allowed to interact with myosin or its subfragments *in vitro*, they are 'decorated' by the myosin molecules which arrange themselves so that all point in the same direction, the so-called arrow head formation. The polymerisation of actin is a dynamic process and in many non-muscle tissues it may exist in the monomeric form which has to be converted to F-actin before contraction can occur. As a consequence of the polarity of the F-actin filament the two ends show some differences in properties. They are designated the barbed, or plus, and the pointed, or minus, ends, the latter being so called as it is the end to which the myosin molecules point when bound. Actin monomers will bind at either end of the filament but as the barbed or 'plus' end has a higher affinity for actin under physiological conditions faster growth occurs at this end (Wegner, 1985). In the myofibril of striated muscle, however, the actin is maintained polymerised as a filament of fixed length, the thin or I filament, where it represents 22% of the total protein and which when examined by electron microscopy looks very like a filament of isolated F-actin. In the myofibril of mammalian muscles the actin filaments are all of identical length, usually about 1.05 μm, the precise length being characteristic for the species. Within the myofibril the actin filament does not appear to be in a dynamic state as is the case with some other cell types. This is possibly a consequence of being anchored at the barbed end by the actin-binding protein, α-actinin, located in the Z band.

An important step to the understanding of the transduction process has been the recent determination of the three-dimensional structure of G-actin by Kabsch *et al.* (1990). Satisfactory crystals for X-ray analysis have not yet been obtained with actin alone, and the structure determination has been carried out on crystals of the complex of actin with DNase I and profilin (Fig. 2.3). Interaction with these proteins no doubt imposes some constraint on the molecules but it is considered that the detailed structure in the complex is close to that taken up in the myofibril. The monomer molecule consists of a large and small

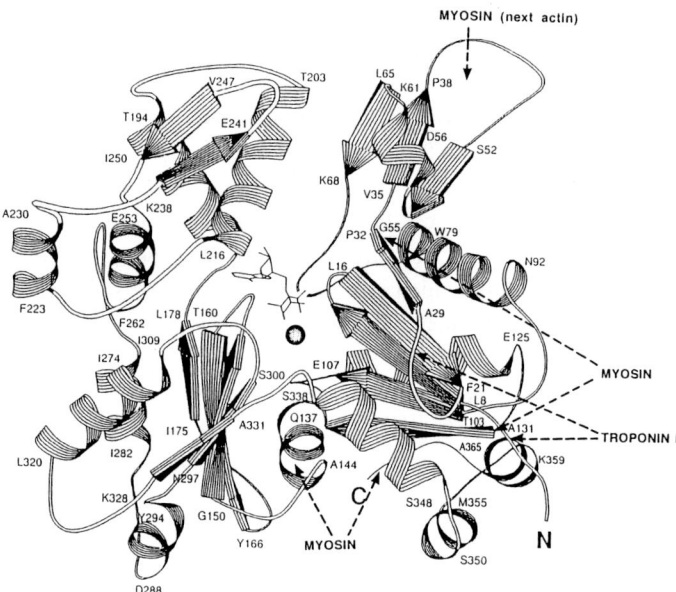

Fig. 2.3. Structure of G-actin determined by X-ray crystallography of the actin–DNase I complex (Kabsch *et al.*, 1990). Arrows indicate presumed regions of interaction of myosin and troponin I with the actin monomers in the thin filament of the myofibril. Note the ADP molecule and calcium ion (hatched circle) located in the base of the cleft.

domain separated by a cleft in which is located the bound ATP. Using the atomic structure of the monomer a model of the F-actin filament has been obtained by fitting the data to the observed X-ray diffraction pattern of oriented F-actin (Holmes *et al.*, 1990). The model (Fig. 2.2(*b*)) indicates that much of the small domain, which includes the N and C terminals, is exposed to solvent and therefore of special functional importance as it includes regions for which there is evidence for interaction with myosin and other proteins of the myofibril.

Actin exists in isoforms that are specific for the muscle type as is the case with other myofibrillar proteins. Nevertheless, the molecule is strongly conserved suggesting that unlike other myofibrillar proteins its properties, in so far as they are controlled by the amino acid sequence, are little modified in different muscle types. In the rabbit the fast and slow twitch muscle actins are identical but different isoforms are present in heart and smooth muscle. The amino acid sequences of the heart and smooth muscle isoforms from the cow only differ in five conservative replacements from the 375 amino acid sequence of the rabbit skeletal muscle isoform. This suggests that the role of actin in the mechanochemical process remains virtually unchanged despite the differences in physiological properties that exist between the muscle types.

Myosin

The protein motor of muscle, myosin, is a large and complex molecule that aggregates to form a bipolar linear polymer, the thick or A filament of the myofibril. As such it is the major component of the myofibril representing about 53% of the total protein. It is an asymmetric molecule of the dimensions indicated in Fig. 2.4 with a molecular mass of about 520 kDa. Myosin is a very polar molecule in that the two ends of the molecule differ markedly in structure and function. The so-called tail is rod-like and designed for lateral association with other myosin molecules to form the A filament. The head of the molecule is most unusual in that it consists of two globular or pear-shaped structures representing the dynamic part of the molecule. The transduction mechanism resides in this region as the pear-shaped heads can be severed from the myosin molecule by controlled proteolytic cleavage to give the preparation known as subfragment 1 (Mueller & Perry, 1962). This protein of molecular mass approximately 130 kDa, with the two light chains intact, retains the contractile properties of the intact myosin molecule.

Myosin is a hexamer built up from three different polypeptides. Two identical polypeptides known as the heavy chains of about 2000 residues make up the bulk of the molecule. The C-terminal portion of the heavy chains are largely α-helical and form a coiled coil in the tail of the myosin molecule. In the N-terminal region the heavy chains separate to terminate separately to form the globular heads where the secondary structure is much less well oriented than in the tail of the molecule. Each head which corresponds to subfragment 1 also

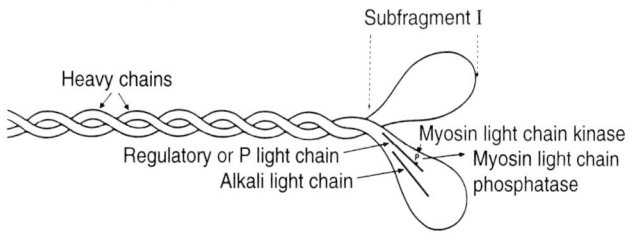

Fig. 2.4. Diagrammatic representation of the myosin molecule. The heads of the molecule, subfragments 1, are 16.5 nm long whereas the tail which is largely α-helical is about 150 nm. The light chains, which are only illustrated on one head, are present in both associated with the neck region of the head which consists of an α-helix of length 8.5 nm at the C-terminus of that part of the heavy chain located in the head (see text and Plate 2.1).

contains two different types of light chains, with molecular masses in the range of 16 to 22 kDa. Although there is evidence that both types of light chain may have roles in the mechanochemical process (see later) their individual functions in vertebrate striated muscle are still uncertain. They are readily distinguished by the fact that one type which is more readily extracted from myosin is phosphorylated by myosin light chain kinase and is known as the regulatory or phosphorylatable light chain. The other type is referred to as the essential or alkali light chain, the latter name reflecting the more extreme conditions required for its extraction from myosin. Often when examined by gel electrophoresis the light chain pattern is more complex than would be expected by the presence of two types. This arises because both types occur in isoforms characteristic of the muscle cell type. Nevertheless each myosin head contains one light chain of each type irrespective of the isoforms present.

At low ionic strength and under conditions that correspond to those existing in the cell, myosin from striated muscle polymerises to form a characteristic bipolar filament structure. The appearance of these filaments in the electron microscope is very similar to the A filament observed in the intact myofibril. It arises from a staggered lateral association of the tails of the myosin molecules with the double heads protruding from the surface of the filament and available for interaction with adjacent F-actin filaments (Fig. 1.1). The central region is devoid of heads and corresponds to the H space in the resting myofibril because the myosin molecules are arranged in opposite directions in the two halves of the filament. The size of the myosin filaments reconstituted *in vitro* are somewhat variable and depend on the conditions. On the other hand, the A filament of the myofibril in a given striated muscle cell type has fixed dimensions containing a constant number of myosin molecules.

Like all protein motors functioning on a filament track

myosin is an ATPase, hydrolysing ATP to ADP and inorganic phosphate. The enzymic activity is located in each of the two heads of the molecule. They may modulate the ATPase but the light chains are not essential for ATP hydrolysis as it is possible to remove both and maintain interaction with actin and enzymic activity, albeit with some modification of the enzymic activity. Nevertheless, they appear to be essential for the mechanochemical process. The sliding velocity of subfragment 1 in model motile systems (see below) is reduced by 90% in the absence of both light chains (Lowey *et al.*, 1993). The removal of one class of light chain gave sliding velocities intermediate between that obtained with the fully deficient and the normal forms of subfragment 1. Quite apart from any enzymic effects the light chains may have, the ATPase of myosin has unique features that reflect its role in the transduction process. Unlike many enzyme systems using ATP as a substrate magnesium is a poor activator of its hydrolysis by myosin alone. In the presence of actin MgATP becomes an effective substrate, a property that is of fundamental significance for muscle.

Myosin exists in a surprisingly large number of isoforms in a given species. The specific isoforms of the heavy and light chains are associated with different muscle cell types and various stages of development. In some apparently mature muscle cells, as is the case with other myofibrillar proteins, more than one isoform of myosin is present. In part this may be a consequence of the fact that each of the three constituent polypeptides is under individual genetic control. Nevertheless, the diversity does have physiological significance for there is a good correlation between muscle speed and V_{max} of the myosin ATPase (Barany, 1967). Despite the sequence differences responsible for the various isoforms the appearance of the myosin molecule in the electron microscope is unchanged.

Organisation of actin and myosin in the myofibril

By the application of electron microscopy and X-ray diffraction the two filament structure of the myofibril can be described with some precision. X-ray diffraction has been of particular value in determining the filament structure because of the precise orientation of the proteins in the myofibril. It has the advantage that measurements can be made on the intact muscle and therefore do not suffer from the possible errors arising from fixation. Much of the work has been carried out on frog sartorius muscle but the evidence available suggests that this is typical for vertebrate muscle and the data correlate well with that obtained with rabbit muscle, on which tissue most of the biochemical work has been carried out. In vertebrate striated muscle the myosin filament is 1.65 μm long, 15 nm in diameter and built up by the lateral, staggered association of about 300 myosin molecules. The packing is such that the heads of the myosin project from the surface of the filament and lie on a three-stranded helical arrangement (Fig. 1.1). The axial repeat is 14.3 nm and the pitch of the helix is 42.9 nm. The myosin filaments lie at the points of a hexagonal lattice and are aligned so that the smooth regions from which heads do not project are in register in the H space of the myofibril.

The actin filaments of length 1.05 μm and diameter 8 nm are attached at the barbed end to each side of the Z line. They occupy the I band and extend up to the boundary of the H space in the A band where they occupy the trigonal positions of the myosin lattice. In the I band the actin filaments are much less well oriented. It follows from this arrangement that the ratio of actin to myosin filaments is 2:1 in vertebrate striated muscle (Fig. 1.1). This ratio varies between different muscle types. In insect flight muscle the actin occupies the diad position of the myosin lattice giving a ratio of 3:1. Higher ratios are found in some invertebrate muscles where the actin filaments are abundant enough to form rings around the myosin filaments.

Whereas the lattice spacing can vary in a given muscle under different conditions and between muscle types, the distance between the surface of the myosin filament backbone and the surface of the actin filament is approximately the same in all muscles, 8–15 nm. This distance presumably represents the limits within which takes place the interaction between the myosin head and the actin filament that is responsible for the transduction process.

Interaction of actin with myosin and ATP hydrolysis

The ATPase activity of myosin is vital for its function as is the case with the other protein motors in eucaryotic cells. It has been pointed out that the magnesium complex of ATP is a poor substrate for the purified enzyme with a very low rate constant at physiological conditions. In the absence of magnesium, however, the CaATP complex is a very effective substrate for myosin ATPase *in vitro*. The latter property is of little significance for muscle where the role of the calcium ion is to trigger off the activation process rather than complex with the substrate. In the living tissue, even during contraction when the intracellular calcium concentration is at its highest, the magnesium concentration is about 100 times greater than that of the calcium. It follows, therefore, that virtually all the ATP in living muscle, whether resting or contracting, is present as the magnesium complex. The capacity of the myosin filament alone to carry out the hydrolysis of MgATP, which corresponds to the situation in resting muscle, is very low. The mechanism of myosin-catalysed hydrolysis of ATP has been studied in detail by a number of groups who have shown that the process is complex and involves several intermediates. The probable sequence of events is outlined below where different conformational forms of myosin (M) are indicated with asterisks.

$$M+ATP \rightleftharpoons M.ATP \rightleftharpoons M^*.ATP \rightleftharpoons M^{**}.ADP.P \rightleftharpoons M^*.ADP.P \rightleftharpoons M^*.ADP + P \rightleftharpoons M + ADP + P$$
$$\quad 1 \qquad\quad 2 \qquad\quad 3 \qquad\qquad 4 \qquad\qquad 5 \qquad\qquad 6$$

A feature of myosin ATPase is that the cleavage step, reaction 3, is much more rapid than the overall turnover of ATP. The limiting step is the process, reaction 5, that results in the liberation of inorganic phosphate (P) and which is very slow with myosin alone. In the presence of actin this step is accelerated and results in the activation of the MgATPase, the rate constant increasing from 0.1 s^{-1} to 20 s^{-1} and higher. The sequence of events involving the interaction between actin and the myosin head that accompany the hydrolysis of ATP by the myofibril and resulting in relative movement between the two filaments is known as the cross-bridge cycle. This is illustrated by the original scheme of Lymn & Taylor (Fig. 2.5) which postulates that in resting muscle the ATPase sites on the heads of the myosin molecules are saturated with the products of ATP hydrolysis. The overall rate of ATPase is low because of the very slow rate of loss of ADP and inorganic phosphate from the myosin enzymic site. When muscle is stimulated the consequent rise in calcium concentration leads to a change in the nature of the interaction of the myosin heads with actin in the neighbouring filaments. This results in a large increase in the rate constant for reaction 5 and the overall rate of ATP hydrolysis. The loss of products from the myosin site is accompanied by conformational changes that lead to relative movement between the two filaments. Once the products are lost MgATP reoccupies the enzymic site, actin dissociates from the myosin and the ATP is rapidly cleaved to ADP.P. If stimulation continues, that is, the calcium concentration remains high, the cycle will be repeated. The scheme illustrated in Fig. 2.5 probably is a good simple representation of the transduction process but much has yet to be learnt about the precise nature of the interaction between the two proteins and the changes, presumably in the myosin head, that accompany ATP hydrolysis and result in the relative movement between the two protein filaments.

It has been apparent for a long time (Perry, 1989) that there are subtle aspects to the interaction between actin and myosin and these are not incorporated in the Lymn–Taylor scheme. In *in-vitro* systems the actin activation of the MgATPase of myosin is very sensitive to increasing ionic strength. Evidence of another type of interaction is given by the formation of a viscous complex between the two proteins that is stable at high ionic

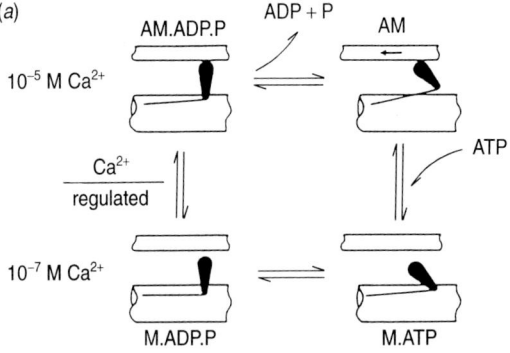

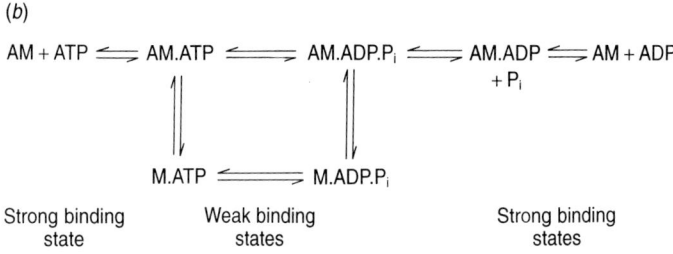

Fig. 2.5. (*a*) Cross-bridge cycle as proposed by Lymn & Taylor (1971). (*b*) Simplified representation of kinetic events occurring during the hydrolysis of ATP by actomyosin (AM). M, myosin.

strengths (more than 0.6) under which conditions activation of the MgATPase is barely detectable or absent. This complex is dissociated at low concentrations of ATP and pyrophosphate. By replacing myosin with its soluble proteolytic fragments actomyosin complexes can be formed and dissociated at low ionic strengths. Despite evidence of dissociation, as judged by viscosity and light scattering determinations on the system, MgATPase remains high. Such studies suggest that the two types of interaction can occur independently and probably involve different sites. The two types of interaction observed *in vitro* by their sensitivity to ionic strength are probably merely reflections of the subtleties of the actomyosin interaction that is at the heart of the transduction process and which occurs at a relatively constant ionic strength. Nevertheless, it is of interest that more recently it has been proposed on the basis of kinetic evidence that weak and strongly bound actomyosin states exist during the cross-bridge cycle (for a review, see Geeves, 1991; Fig. 2.5(*b*)). Implicit in kinetic schemes incorporating two states of interaction is the presumption that even in resting muscle the myosin head is weakly attached to actin. In the myofibril of resting vertebrate striated muscle this does not lead to activation of the MgATPase because of the action of the regulatory proteins also present in the thin filament (Chapter 4). These proteins prevent the weak interaction of actin with myosin observed *in vitro* with the purified proteins that leads to activation of the MgATPase. Thus it follows that the latter weak interaction cannot be identical with that postulated to occur in resting muscle in the kinetic scheme, although there may be some common aspects.

On stimulation the ATPase level rises and it can be postulated that the weak interaction is supplemented by (or replaced with) the formation of the strongly bound (rigor) state with the result that movement between filaments takes place. This implies that the strong interaction that is responsible for forming the actomyosin complex *in vitro* at high ionic strength plays a role in developing tension in the myofibril on contraction. If this assumption is correct the fact that *in vitro* interaction only occurs in the absence of ATP would suggest that it takes place at that point in the cross-bridge cycle when the nucleotide momentarily is not bound to myosin and the molecule is in a conformation that enables it to form a strong interaction with actin.

To understand its subtleties it is necessary to be able to describe the actin–myosin interaction in detailed structural terms. The prospects for this are now good with the recent determination of the three-dimensional structure of actin and subfragment 1 at high resolution.

Cross-bridge structure and function

The concept of the cross-bridge arose from the earlier high resolution electron micrographs of vertebrate skeletal muscle that showed marked bridge-like structures extending between the myosin and actin filaments. These were identified with the heads of the myosin molecules where the ATPase and actin-binding properties are known to reside. In rigor muscle, particularly that from the flight muscles of certain higher insects, such as *Lethocerus maximus*, the cross-bridges attach to actin filament at an angle of 45°. This is similar to the attachment of isolated myosin or subfragment 1 molecules to F-actin filaments and was considered to represent the position of the myosin head at the end of the power stroke. As illustrated in the scheme in Fig. 2.5 relative movement was considered to take place by the movement of the myosin head attached to an actin monomer in the filament from the 90° position occupied before activation to the 45° position. The driving force for this change would be provided by the hydrolysis of ATP. Nevertheless, methods designed to measure the angle of attachment of the myosin head during contraction suggest that it does not undergo a change of the order demanded by a mechanism of force generation involving rotation of the myosin head (for reviews, see Squire, 1988; Vibert & Cohen, 1988). Likewise there is no evidence for a marked change in conformation of the myosin head because neutron scattering studies indicate that the difference in shape between the initial and final stages in the contractile cycle is not greater than 20%. There have been difficulties in demonstrating a change in the angle of attachment of the myosin head to the actin filament but recent fluorescence polarisation studies with glycerated fibres have shown that the regulatory light chain region of the myosin head tilts during imposed filament sliding and during the subsequent quick force recovery. Assuming all the heads are involved in the length step, however, the amount of tilt is less than 3°, much smaller than the conventional contractile models demand (Irving *et al.*, 1995).

X-ray diffraction studies on living muscle support in general a cross-bridge mechanism of contraction although the resolution is not adequate to define the changes in conformation that occur to produce relative movement of the filaments. They indicate that there is a movement of mass from the A to the I filament which is not observed if the muscle is stretched so that the filaments do not overlap. The helical arrangement of the myosin molecules in the A filament, which is responsible for the 143 Å periodicity, does not change during contraction and neither do the pitch and subunit repeat of the actin filament (Haselgrove, 1983). Such studies demonstrate clearly that no marked changes occur in the axial periodicity of either of the two filaments during contraction. The movement of material from the A filament to the I filament indicated by the changes in the diffraction pattern is taken to represent the myosin head attaching to the actin subunit. In the frog sartorius muscle this process is complete in 25 ms and tension develops in 40 ms, suggesting a substantial number of cross-bridges may form without developing tension. The intensity of the X-ray reflections during the steady active state also suggests that only a fraction of the cross-bridges are attached in an ordered manner.

Model motile systems

The development in recent years of model motile systems using purified F-actin and myosin in which one of the components is immobilised either in a filament or attached to a surface has enabled the transduction process to be studied in more detail (for reviews, see Huxley, 1990; Burton, 1992; Spudich, 1994). Early work which hinted at the possibility of observing the relative movement of individual actin and myosin molecules was that of Oplatka & Tirosh (1973). These authors observed active streaming in actomyosin solutions which they interpreted as a consequence of the interaction between the actin and myosin molecules in solution. More recently, it has been demonstrated that myosin and its subfragments possessing the head move along the F-actin filament with a speed that is independent of the filament length in a direction determined by the actin, that is, towards the barbed end as is the case in the intact myofibril. In contrast, actin filaments will move in either direction along a linear track of heavy meromyosin molecules (Toyoshima *et al.*, 1989), the direction in this case being determined by the orientation of the actin filaments. This indicates that the myosin head is attached with sufficient flexibility to adjust its position with respect to the tail of the molecule to accommodate the polarity of the adjacent actin filament. A particularly important finding (Toyoshima *et al.*, 1987) is that subfragment 1 is able to move along actin filaments. This clearly indicates that the mechanism for transduction resides in the head alone consisting of 840 amino acid residues in the N-terminal region of the myosin heavy chain. These observations are incompatible with models proposed for cross-bridge movement that merely involve hinge-like regions in the tail or at the head–tail junction.

Plate 2.1

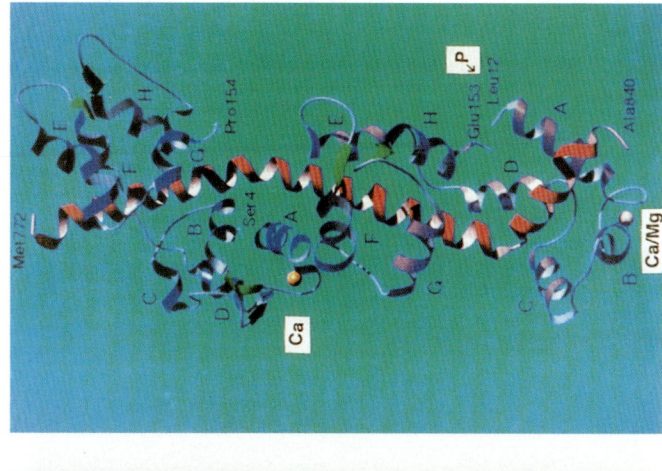

(b)

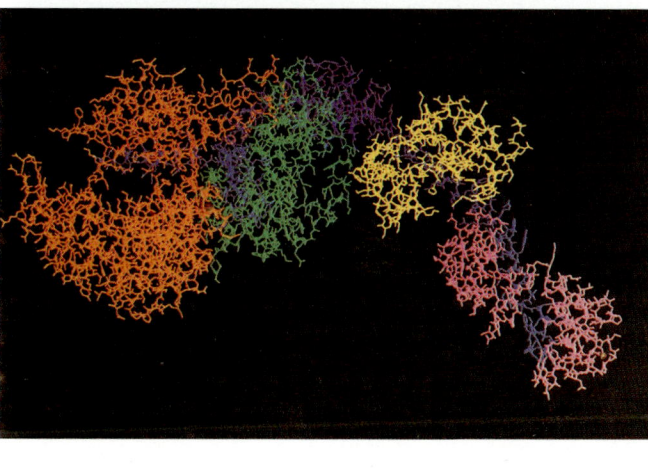

(a)

(a) Structure of myosin subfragment 1 as determined by X-ray crystallography of crystals of the reductively methylated form of the protein. The 50 kDa tryptic fragment (orange) is at the top with the regulatory (phosphorylatable) light chain (magenta) and the essential light chain (yellow) bound round the helical region of the 20 kDa fragment (blue) at the lower part of the figure. The small sphere on the regulatory chain (lower left) marks the divalent metal binding site. The ATPase active site occurs at the interface of the N-terminal 23 kDa fragment (green) and 50 kDa and 20 kDa fragments. The actin-binding site is at the upper right (orange) over the large cleft and near the start of the 20 kDa fragment (blue). Reproduced from Yount (1993) after Rayment et al. (1993a).

(b) Ribbon diagram of the regulatory domain of scallop myosin. The helical domain of the subfragment 1 is shown in red and the essential and regulatory light chains in blue and pink, respectively (from Xie et al., (1994) The overall structure exhibits similarities to the corresponding region of subfragment 1 of chicken skeletal muscle but differs in that there is a novel calcium binding site in the essential light chain that is stabilised by linkages involving the heavy chain and both light chains. Regulation of the actin-activated ATPase of scallop myosin involves the binding of calcium at this site. The non-specific divalent cation binding site is indicated at the lower end of the regulatory domain. There is no clear evidence that vertebrate smooth muscle myosin is regulated by calcium binding at such a unique site. It is probable that vertebrate smooth muscle myosin possesses a similar structure although actin activation is regulated by phosphorylation of the site on the regulatory light chain corresponding to that indicated on the lower right of the diagram.

Measurements made on these molecular motile systems have enabled certain parameters of the contractile process to be determined. It has long been considered on the basis of a range of evidence obtained from whole muscle that the length of the working stroke, that is, the distance moved in one cycle involving the hydrolysis of one molecule of ATP as illustrated in Fig. 2.6, is 10–20 nm. Using the model systems Spudich and collaborators originally reported values in the range of 5–30 nm for the working stroke, which fit in well with the earlier estimates. On the other hand, Yanagida *et al.* (1995) concluded from their data that the distance of sliding per molecule of ATP hydrolysed is more than 60 nm to less than 200 nm.

The reasons for this large discrepancy between the investigators probably relate to the methodologies employed by the different groups. In the original studies, to obtain values for the working stroke it was necessary to measure the velocities of short actin filaments moving over a surface on which the myosin heads are bound, the length of the actin filaments and the rate of ATP hydrolysed. Assumptions have to be made as to the number of myosin heads attached to the short filaments at any one time but even so it is surprising that the values reported differ so much. Recent advances in techniques which avoid some of the assumptions referred to above have enabled the direct measurement of force and the displacement of single myosin molecules. Studies of this type using optical tweezers suggest that the step size may be closer to the lower values reported. With this technique Finer *et al.* (1994) observed discrete stepwise movements of 11 nm. When corrections for Brownian movement are applied to the optical tweezer technique a lower value of 5 nm has been reported (Molloy *et al.*, 1995). In view the wide range of values reported it may be necessary to consider the possibility that the step size depends on the precise conditions under which the muscle is working. Ishijima & Yanagida (1995) in studies on their motile model system of actin and myosin filaments distinguish the power stroke, which they determine as 17 nm, from the myosin step size, the sliding distance associated with the hydrolyis of one molecule of ATP. According to their model, in the tight coupling state one power stroke of 17 nm accompanies one ATPase cycle. In the loose coupling state several power strokes can accompany one ATPase cycle, giving a myosin step size of about 100 nm.

Under isometric conditions of low load single force transients of 3–4 pN were obtained by Finer *et al.* (1994). A similar value for the force developed by a myosin head can be derived assuming that 50% of the free energy of ATP hydrolysis is available to perform work and taking the experimental value of 50–60% for the efficiency of energy of conversion obtained with frog sartorius muscle. If one molecule of ATP is available per cross-bridge cycle with a working stroke of 15 nm the force developed would be at least 3 pN (Woledge, 1988). These values are significantly higher than those obtained experimentally for the force developed per cross-bridge in whole muscle, about 1 N. In view of these differences Finer *et al.* (1994) conclude that 20–40% of the cross-bridges develop tension at any one time. This conclusion is similar to that arrived at from measurements of the intensity of layer lines observed in X-ray diffraction studies of contracting frog muscle. They suggest that only 20–30% of the myosin heads are in the ordered bound state (Huxley & Kress, 1985). On the other hand, Kishino & Yanagida, (1988), calculate that the force developed per cross-bridge when an actin filament interacts with myosin or subfragment 1 bound to a glass plate in the presence of ATP is 0.4–0.6 pN. These values would appear to be too low for they require that in contracting muscle all myosin heads are developing tension whereas the weight of evidence from a number of techniques is that at one time only a fraction are so engaged.

In models of the cross-bridge cycle it is generally assumed that the cycle is tightly coupled to the hydrolysis of one molecule of ATP. Much of the experimental data is compatible

with this view, that is, that a movement of the myosin head of
about 10–15 nm accompanies the hydrolysis of one molecule of
ATP. Some investigators, particularly Yanagida and collaborators, have described experiments suggesting that larger dis-
placements can accompany the hydrolysis of a single molecule
of nucleotide. Assuming that the working stroke is now established at 10–15 nm this would imply that a number of steps of
this size can accompany the hydrolysis of one molecule of ATP.
It is difficult to visualise the mechanism of such a process but
while the discrepancies in the experimental data persist it would
be wise not to dismiss it out of hand. There are considerable
technical difficulties in measuring the amount of ATP hydrolysed during a single cross-bridge cycle by a single myosin
molecule. Until this can be achieved this question cannot be
finally answered.

Structure of the myosin head

The myosin molecule is unusual in that it contains two identical
heads each of which possess actin-binding and ATPase activities and which individually, as subfragment 1 molecules, are
able in association with F-actin to carry out the transduction
process. The reason why the functional molecule in the A
filament is double-headed is far from clear. It will certainly
increase the capacity of the A filament to form cross-bridges and
the double helical tail will exhibit greater rigidity than would be
the case with a myosin molecule with one head and a single
α-helix of the heavy chain in the tail. This rigidity may be
important for the formation of an A filament with the properties
required for its function in the myofibril.

Electron microscope studies have indicated that the myosin
head is elongated and pear-shaped. The recent three dimen-
sional structure to 2.8 Å resolution obtained by X-ray diffraction studies on crystals of subfragment 1 (Rayment *et al.*, 1993*a*)
in general confirm the earlier views on the shape of the molecule.

The myosin head is highly asymmetric with a length of 165 Å, width 65 Å and thickness of approximately 40 Å. To explain the mode of action of the myosin head it is necessary, in addition to understanding the conformation of the heavy chain, to define the sites of interaction of the various ligands and proteins associated with this region of the molecule. Analysis of high resolution electron micrographs has provided limited information on the location of the interacting molecules on the myosin head, information which is vital for the understanding of the transduction process. Much information, obtained by a variety of biochemical and biophysical techniques, regarding the amino acid residues of the primary heavy chain sequence that are involved at the various interaction sites has existed for some time (for a review, see Mornet *et al.*, 1989). The determination of the three-dimensional structure by Rayment *et al.* (1993*a*) has extended this information and enabled it to be related to the three-dimensional structure of the head (Plate 2.1(*a*)).

Sites susceptible to proteolytic digestion have long been described in the heavy chain of subfragment 1. By digestion with trypsin under defined conditions three fragments of molecular masses 20, 27 and 50 kDa corresponding to segments of the myosin head have been isolated. Actin binds to the central 50 kDa segment and to the carboxy terminal 20 kDa segment. These fragments were originally considered to be functional domains but this now does not appear to be the case. Their formation merely reflects the position of residues in the heavy chain that are susceptible to proteolysis.

Approximately 48% of the amino acid residues are in the α-helical conformation, a striking feature of which is an 85 Å run of α-helix extending from the thickest part of the head to the carboxy terminus of the heavy chain (Plate 2.1(*a*)). The regulatory (phosphorylatable) light chain is wrapped around the end of the α-helix distal from the actin and nucleotide-binding sites. This light chain, which bears considerable structural homology to calmodulin and troponin C, consists of two domains. The

N-terminal domain interacts with the heavy chain between residues asparagine 825 and leucine 842 and the C-terminal domain in the region bounded by glutamic 808 and valine 826. The alkali essential light chain is wrapped around the part of the large α-helical run interacting in the region of leucine 783 to methionine 806. In effect the two light chains, which do not overlap to any significant degree, increase the mass of the neck region of the myosin head where it attaches to the double helical chain of the molecule. As this region of the heavy chain consists of a single α-helical chain the presence of the heavy chains might be expected to increase its rigidity. If this is the case and a rigid helical neck domain is essential for the mechanochemical process the much reduced velocity of myosin head movement observed in motility assays in the absence of light chains can be explained.

The thick part of the myosin head contains the 50 and 27 kDa segments and part of the 20 kDa segment. It is composed mainly of a large, mostly parallel, seven-stranded β-sheet with associated α-helices and loops. This part of the molecule is of particular importance for the function of myosin because it contains the nucleotide and phosphate-binding sites associated with the active site for the ATPase and the region of the molecule to which actin binds. The region of the polypeptide chain that gives rise to the 20 kDa fragment extends along the length of the molecule and contains the N-terminal α-helix to which the light chains are bound and contributes to the ATP and actin-binding sites.

Although the structure was determined on crystals from which ATP was absent it has been possible to identify the nucleotide-binding site from the positions of residues at the site previously identified by the use of photosensitive ATP analogues (Yount *et al.*, 1992) and by analogy to the phosphate-binding loops in the Ras protein and adenylate kinase. The residues involved in binding MgATP are contributed by all three proteolytic fragments. In the crystals a sulphate ion is bound close to serine 181 located in a conserved sequence in the 23 kDa fragment and to serine 243 from a conserved sequence in the 50 kDa fragment. This is considered to be the γ-phosphate-binding site of MgATP (Yount, 1993). It is likely that the site which is 13 Å deep and 13 Å wide with an angle of 40° between the faces, closes when ATP is bound. Myosin has long been known to possess reactive sulphydryl groups (SH1 and SH2) at cysteine 707 and cysteine 697, respectively, that are important for its enzymic activity. These residues lie on opposite sides of two helices in this region and can only be cross-linked when nucleotide is bound, indicating a large change in distance between the two cysteines that would be compatible with the closing of the active site. The conformational changes that arise from closing of the site on ATP binding could be of great functional significance in the transduction process. Rayment *et al.* (1993*a*) suggest that if the actin-binding face of the myosin head remains stationary, closure of the nucleotide-binding cleft could produce a movement of the carboxyl terminus of the α-helical neck region of approximately 60 Å. This value is getting close to that taken to be the size of the working stroke.

Structural aspects of the actin–myosin interaction

Crystals of the subfragment 1–actin complex which would enable the interaction sites to be defined with precision by X-ray crystallography are not yet available. Nevertheless, labelling and other studies (for reviews, see Mornet *et al.*, 1989; Levine *et al.*, 1990) have indicated that a region including the junction of the 50 kDa and 20 kDa segments interacts with the N-terminal region of actin and the alkali light chain with the actin C-terminus. The information currently available indicates that the actin-binding site is about 50 Å distant from the ATPase site and on the opposite side of the head. It has long been known that when ATP or pyrophosphate binds to myosin, in solution at least, actin dissociates implying that the binding of ATP at the active site produces structural changes that are in some way conveyed to the actin-binding region, changing the nature of the interaction there. The passage of information between these sites is two-way. In the reverse situation when actin is allowed to interact with myosin in the presence of ATP the environment at the nucleotide-binding site is changed so the rate of MgATP hydrolysis increases by two orders of magnitude. When these changes occur in the gel state as exists in the myofibril they probably correspond to discrete steps in the cross-bridge cycle. For energy transduction the interaction between actin and myosin must be dynamic, passing through a series of stages and returning to the original resting state so that the cycle can be repeated.

Structural information derived from X-ray diffraction studies on protein crystals of subfragment 1 and actin and image reconstructions from electron micrographs will relate to conformations into which the molecules are locked as a consequence of the method of preparation of the samples. In this respect consideration must be given to the fact that it was necessary to subject subfragment 1 to reductive methylation to produce crystals suitable for high resolution structural studies. This procedure involves modification of lysines, most of which lie on the surface of the molecule. Similar modification of lysozyme does not change its structure determined at 1.8 Å resolution. Whether the same applies to subfragment 1, a much larger molecule, which because of the nature of its function would be expected to have the capacity to undergo pronounced conformational changes, is uncertain. The methylated subfragment 1 has ATPase activity with some of the kinetic parameters modified; for example the actin activation of the MgATPase is less than one-tenth of the effect on the unmodified protein (White & Rayment, 1993). Nevertheless Rayment *et al.* (1993*a*) do not consider that there are major structural changes from the native form of the protein. Discounting any changes which may arise from methylation one cannot be sure as to which particular stage in the cross-bridge cycle that the conformation examined corresponds. It would seem probable that the structure studied by Rayment *et al.* (1993*a*) in the absence of ATP corresponds to that existing in the rigor state in which myosin forms a strong link with actin.

The nature of the links between actin and myosin and the transformations that they can undergo is vital for understanding the transduction process. An approach to this problem is provided by the availability of the high resolution structures of both the myosin head and actin. Combining this information with the image reconstruction of electron density maps of F-actin and F-actin decorated with subfragment 1 has enabled a structure for the actomyosin complex to be modelled (Rayment *et al.*, 1993*b*).

This analysis concludes that the binding site domain on myosin binds tangentially to the actin filament at an angle of about 45° to the actin filament axis. The helical tail of the myosin head where the light chains are bound projects away at an angle of about 90° to the actin filament axis. A number of regions where the amino acid side chains of the two proteins are sufficiently close to imply that hydrophobic or ionic interactions could occur can be identified from the model. The model indicates, as has long been suggested (Perry & Cotterill, 1965) that each myosin head interacts with two adjacent actin monomers, involving the primary and secondary binding sites. It is suggested that probably the first event in the docking of actin on the myosin head involves an interaction in the region extending between tyrosine 626 and glutamine 647 which includes a lysine-rich loop and connects the 50 and 20 kDa segments. As this loop is close to the N-terminal region of actin that is rich in acidic residues an electrostatic interaction occurs. This will be a relatively weak interaction sensitive to ionic strength and Rayment *et al.* (1993*b*) suggest that it would be followed sequentially by: (1) a stereospecific interaction involving hydrophobic residues, and (2) a strengthening of this interaction by the recruitment of additional groups from the upper 50 kDa domain; thus leading to an increase in binding constant as would be required for the power stroke.

The initial weak interaction could be partially responsible for the ionic strength-dependent weak binding characteristic of actomyosin at low ionic strength. Under these conditions MgATP is hydrolysed at the high rate associated with contraction. If the suggestion is correct the interaction must produce effects that are transmitted to the nucleotide binding site, about 50 Å away.

The stereospecific hydrophobic interaction involves two segments in the end of the lower domain of the 50 kDa fragment of the myosin heavy chain extending from residues 516 to 558. In this region the interface involves two adjacent actin monomers, one for each segment of the myosin interface. The third type of interaction involves residues from the upper domain of the 50 kDa segment. One feature of the latter segment, upon which the majority of stereospecific binding sites are located, is a narrow cleft that extends from under the nucleotide-binding pocket to the actomyosin interface. The best fit of the models to the image reconstruction produces a collision between the actin and myosin in this region which can be relieved by closing the cleft. The authors have therefore suggested that the formation of the rigor complex involves closure of this cleft. If this is the case it follows that the crystal structure of the methylated subfragment 1 does not correspond to that of the rigor state even though it was crystallised in the absence of ATP. The concept of opening and closing of the cleft is attractive because it provides a means of communication between the actin and nucleotide-binding sites. It is easy to visualise how the conformational changes that occur at the nucleotide site when ATP is bound could readily be transmitted along the cleft resulting in its opening and hence weakening the binding of actin and ultimately leading to its dissociation from the myosin. Also, as pointed out above, the effects arising from conformational changes when actin binds can also be transmitted to the nucleotide-binding site. The direction in which the effects are transmitted will probably depend on the type of interaction that actin makes with myosin. In the light of these modelling studies Rayment *et al.* (1993*b*) have proposed a mechanism involving opening and closing of the cleft during the contractile cycle. In this scheme, which is illustrated in Fig. 2.6, the distortion of the subfragment 1 tail that occurs when ATP is bound and the nucleotide site closes provides the force that enables relative movements between the filaments to take place.

Despite recent progress much has still to be learnt about the nature of the interaction of actin and myosin and the effect of ATP upon it that leads to the contractile process. The determination of the high resolution structure of the two proteins is a major step forward in the understanding of the transduction problem. The model proposed postulates a large movement of the tail region of the myosin head but most data obtained so far suggest the bulk of the myosin head does not move during cross bridge cycling. Recently cryoelectron microscopy coupled with image analysis indicate that the rigid light chain binding domain of smooth muscle myosin swings 23° (Whittaker *et al.*, 1995) and brush border myosin I, 32° (Jontes *et al.*, 1995) on ADP release. Such swings would correspond to movements at the more distal region of the proton motor of 35 Å and 50 Å, respectively.

Another aspect of the interaction which requires to be explained is how actin is able to influence events at the nucleotide site on the myosin head to activate the hydrolysis of MgATP. This interaction of actin must be the first event in the contractile cycle that could then be followed by the strong rigor-type interaction, the power stroke. The rigor-type interaction could be the consequence of the closure of the cleft in the 50 kDa segment at the moment when the ADP and inorganic phosphate are lost from the nucleotide site. Subsequent binding of ATP to the nucleotide site would lead to opening of the cleft and dissociation of the actin from its strong binding site. Such a scheme which differs slightly from that of Rayment *et al.* (1993*b*) in the order of events requires that in the absence of ATP from the nucleotide-binding site the cleft on the myosin head is closed. One cannot be sure, however, that the methylated subfragment 1 is in the rigor conformation as would be expected

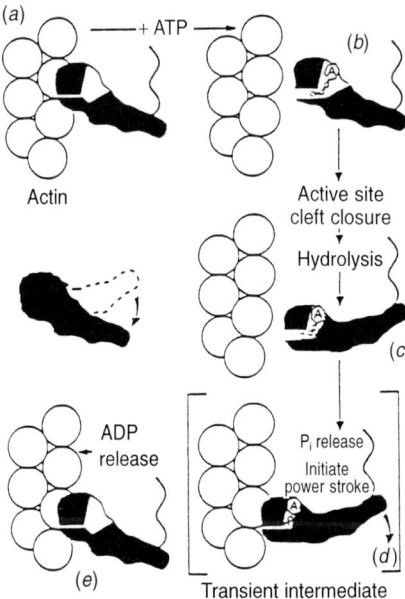

Fig. 2.6. (*a–e*) Scheme for the contractile cycle incorporating the structural features of the head and the results of modelling the actomyosin complex. The narrow cleft that splits the 50 kDa segment is represented as a horizontal gap at a right angle to the filament axis. In the model the cleft is at an angle of about 30° to the axis. The representation of the nucleotide-bound state and the associated conformational change in relation to the X-ray structure of myosin is conceptual in nature. Reproduced from Rayment *et al.* (1993*b*).

in the absence of the nucleotide. An important step forward in resolving this problem will be to determine the structure of subfragment 1 to which the nucleotide is bound at the active site.

In-vitro systems containing actin and myosin alone will hydrolyse ATP until such time as all the substrate is used up. The unique feature of intact muscle is that it can be regulated with great speed and precision by a change in calcium concentration. In all types of muscle this is brought about by modulating the actin effect on myosin that leads to a large increase in ATP hydrolysis. This modulation either involves the regulatory light chain on the myosin head or additional protein systems that interact with the actin itself (Chapters 3 and 4). These aspects must be accommodated in any scheme explaining energy transduction by the actomyosin system in functioning muscle.

3

Calcium and the regulation of contraction

In all types of contractile systems that depend on the interaction of filaments of actin and myosin the molecular events giving rise to contraction are presumably very similar. Many of the differences in properties of these contractile systems will therefore depend on the nature of the regulatory mechanisms involved in their control. Contraction in all muscles is triggered by an increase in the Ca^{2+} concentration to 100–1000 times the resting level. It therefore follows that the differences in contractile response of different muscle types arise mainly from the mechanisms that convey the calcium signal to the actomyosin system. Muscle, by employing a change in Ca^{2+} concentration to initiate contraction, is using a mechanism that triggers a wide variety of cellular events. In comparison with many intracellular responses to external stimuli contraction is an extremely rapid process, usually completed in milliseconds. Muscle contraction involves the response of the whole cytoplasm and is not confined to a specific region of the cell as is sometimes the case with other intracellular responses. For these reasons there is a very large rapid flux of calcium within the cell when muscle contracts. To accommodate this requirement one of the aspects of muscle specialisation is the ability to concentrate calcium rapidly into large intracellular stores that can be mobilised in milliseconds. This process requires the expenditure of significant amounts of energy. It has been estimated that in actively working fast skeletal muscle the chemical energy used to mobilise the calcium required to regulate the process may be as much as a fifth or a quarter of that used to perform mechanical work.

Muscle cells exist in an environment in which the external Ca^{2+} concentration is 10^{-4} to 10^{-3} M or higher. The use of calcium as a signalling system requires that the normal resting intracellular Ca^{2+} concentration be kept low, usually about 10^{-7} M, with mobilisable stores of the cation readily available. The main intracellular stores of calcium are in the endoplasmic reticulum and in the mitochondria. Cells possess powerful calcium pumps to maintain homeostasis and mechanisms to ensure the specific release of the cation from the stores by various agents. The pumping systems can be distinguished by their locations in the cell.

(1) *Surface membrane*
    (a) Mg/Ca activated ATPase mechanistically linked to calcium translocation. Molecular mass of 130 kDa, acylphosphate intermediate, modulated by calmodulin and requires membrane phospholipids.
    (b) Electrogenic exchange of sodium and calcium ions
(2) *Endoplasmic reticulum*
    Mg/Ca activated ATPase with molecular mass of 110 kDa, apparently distinct from the surface membrane enzyme. Relatively insensitive to calmodulin but dependent on phospholipids. The electrogenic transport against a large electrochemical gradient of two calcium ions per molecule of ATP hydrolysed involves transient phosphorylation of the enzyme.
(3) *Mitochondrion*
    Energised by electron transport or ATP hydrolysis. Mechanism and role not clear but system has moderate affinity and probably has storage and buffer functions.

The relative importance of the above systems in maintaining calcium homeostasis depends on the cell type. Surface membrane systems are concerned with the maintenance of the total cell calcium as for example in the erythrocyte and excitable tissues such as heart and nerve. The intracellular distribution of calcium is largely controlled by the endoplasmic reticulum which is particularly abundant in striated muscle tissue where it is known as the sarcoplasmic reticulum (SR). The particular significance of the SR in muscle is indicated by the fact that it is most abundant and specialised in the fastest vertebrate skeletal muscles.

Sarcoplasmic reticulum

The SR is a convoluted structure lying on the surface of the myofibrils and bounded by a membrane (Fig. 3.1). It has two

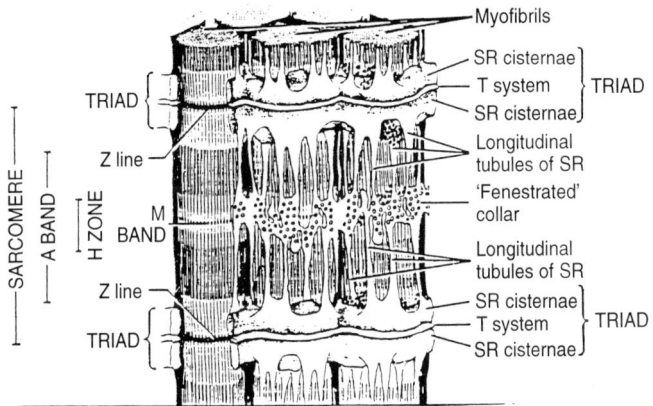

Fig. 3.1. Diagrammatic representation of the sarcoplasmic reticulum of frog sartorius muscle showing its regular arrangement with respect to the myofibrils. In frog sartorius muscle the triads are located at the level of the Z line with the result that there is one triad per sarcomere. Two triads per sarcomere are present in most other animals including humans where they are positioned close to the junction of the A and I bands. Reproduced from Martonosi (1986); modified from Peachey, *J Cell Biol*, 1965 **25**, 209.

main functions: namely (1) calcium transport and storage, and (2) calcium release into the sarcoplasm, each of which are associated with morphologically distinct regions. The bulk of the structure consists of lateral sacs and finger-like projections, the longitudinal tubules, extending along the axis of the myofibril. This is sometimes called the free SR with which most of the calcium transport ATPase is associated, evenly distributed along the membrane. Electron-dense material seen in the lateral sacs (terminal cisternae) probably consists of calsequestrin, the storage protein to which the calcium is bound. The longitudinal tubules contain little or no calsequestrin. The region of the SR lying close to the transverse (T) tubules appears to connect to the latter system through feet-like projections on the surface of the membrane in this region of the terminal cisternae. This region which is known as the junctional SR is concerned with transmission of the excitatory impulse from the T-tubules to the SR causing calcium release. The transverse tubular system with the associated adjacent terminal cisternae is known as a triad (Fig. 3.1).

Calcium transport system

The calcium transport ATPase is a single polypeptide of molecular mass of about 110 kDa (MacLennan *et al.*, 1985). It is a major component of the SR and may represent from 50 to 80% of the total protein. The higher values are found in very fast muscles where the amounts correlate with the capacity to transport calcium. The SR calcium pump can lower the sarcoplasmic calcium concentration to or below 10^{-8} M and maintain a gradient of 1000-fold across the membrane. The storage capacity is approximately 0.2 μmole of calcium per mg of SR protein. In the presence of Ca^{2+} precipitating anions to which the membrane is freely permeable, the calcium concentration can be increased to 7–8 μmole per mg protein. In the luminal space of the SR in resting muscle the calcium concentration may reach millimolar (Martonosi, 1986).

As an intrinsic membrane protein the transport ATPase requires phospholipid for activity. Hydrophobic parts of the molecule are considered to span the membrane with much of the polypeptide chain folded into a globular mass on the sarcoplasmic side. It is considered that only short segments of the molecule are exposed on the luminal side (Fig. 3.2). The stoichiometry of the reaction appears to be well established, namely two calcium ions transported for every molecule of ATP hydrolysed. The reaction is considered to take place by a number of steps.

(1) One molecule of ATP and two calcium ions interact with part of the enzyme on the sarcoplasmic side of the membrane to form a complex.

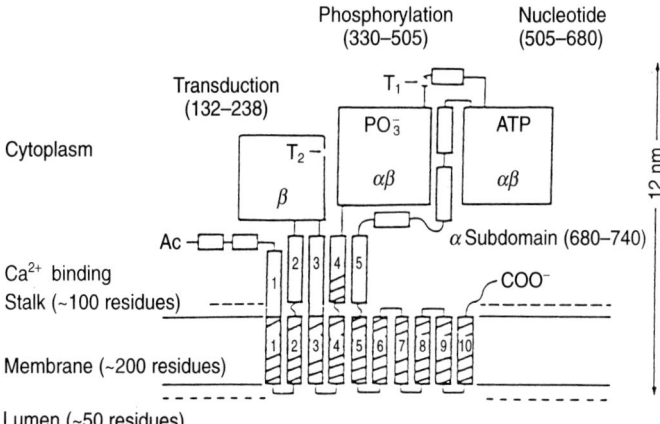

Fig. 3.2. Diagrammatic representation showing the major domains of the calcium transport ATPase in the membrane of the sarcoplasmic reticulum and their functions. The predicted helices would probably form tight clusters. The ATP-binding domain is probably folded alongside the two other major domains to account for the approximately molecular profile observed in electron micrographs of the ordered arrays observed in the membrane of the sarcoplasmic reticulum. The site of phosphorylation is aspartic 351. Reproduced from Tada & Kodama (1989).

(2) The binding of ATP and calcium causes a conformational change in the enzyme, ATP is hydrolysed with the phosphorylation of an aspartyl residue at the active site. The calcium ions are now in some way incorporated into the system so that they become unavailable to ethylene glycol-bis(β-aminoethyl ether) N,N'-tetraacetic acid (EGTA).

(3) The phosphorylated enzyme undergoes conformational change which results in Ca^{2+} being translocated across the membrane.

(4) Ca^{2+} is released into the luminal space. The release of Ca^{2+} is progressively inhibited as the concentration of the cation increases.

(5) The enzyme undergoes isomerisation, dephosphorylation catalysed by Mg^{2+} occurs and phosphate is released on the sarcoplasmic side of the membrane.

The normal function of the enzyme is to concentrate calcium inside the SR, although the system is reversible. In the presence of ADP and inorganic phosphate, high intervesicular and low external Ca^{2+} concentration, ATP is synthesised with the release of two calcium ions into the external medium.

Calcium release

To accommodate the speed of muscle contraction calcium must be released from the internal stores rapidly and at many sites distributed throughout the sarcoplasm so that all parts of the contractile machinery can respond virtually simultaneously. The close association of the SR with each myofibril ensures that the maximum distance that the Ca^{2+} ions have to diffuse to reach their targets does not exceed 2–4 μm. Over the years there has been much interest in the mechanism of calcium release from the terminal cisternae and a number of theories proposed (for reviews, see Martonosi, 1986; Ashley *et al.*,

1991). Recent advances in knowledge of the structure and composition of the junctions between the terminal cisternae and the transverse tubule system, however, have very much illuminated this subject (Figs. 3.3 and 3.4). This region is obviously important in the excitation–contraction sequence of events. As the wave of depolarisation resulting from nerve stimulation sweeps inward along the transverse tubules, it is the region where changes in cell membrane potential could

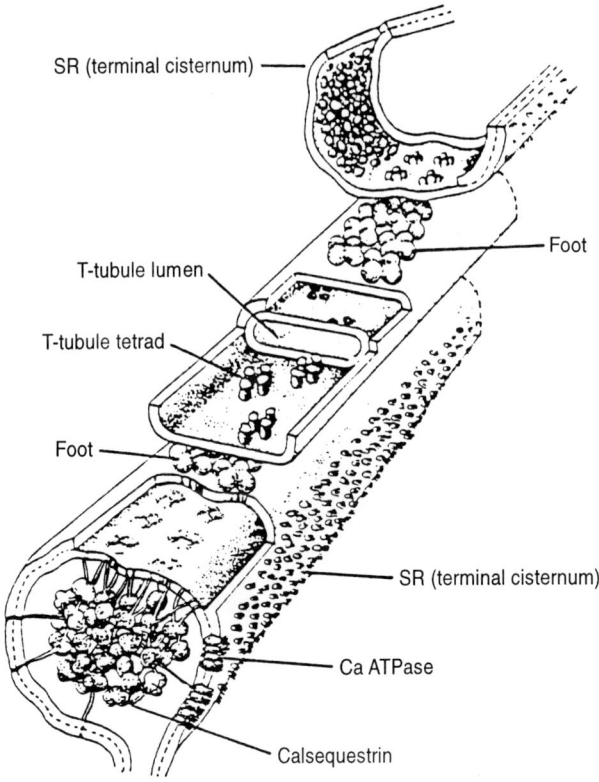

SR (terminal cisternum)

Foot

T-tubule lumen

T-tubule tetrad

Foot

SR (terminal cisternum)

Ca ATPase

Calsequestrin

Fig. 3.3. Diagrammatic representation of a triad showing the relative positions of the foot proteins, T-tubular tetrads, calcium transport ATPase and calsequestrin. Reproduced from Block *et al.* (1988), by copyright permission of The Rockeller University Press.

have a direct effect on the permeability of the membrane of the SR.

For some time globular particles that occupy the space in the junctional region and named 'feet' by Francini-Armstrong (1975), have been considered to play a role in excitation–contraction coupling (Fig. 3.3).

The isolation of the ryanodine receptor and the demonstration by electron microscopy that the isolated receptors resembled the foot protein particles confirmed that this was indeed the case (Fleischer & Inui, 1989; Lai & Meissner, 1989). Ryanodine is a plant alkaloid that at micromolar concentrations depresses twitch tension and causes a slow contracture of the muscle. When the isolated receptor was incorporated into planar lipid bilayers it gave calcium channels with properties similar to those of the terminal cisternae membrane. In freeze–fracture electron micrographs of the SR membrane the receptor appears as a square particle made up of four subunits. More

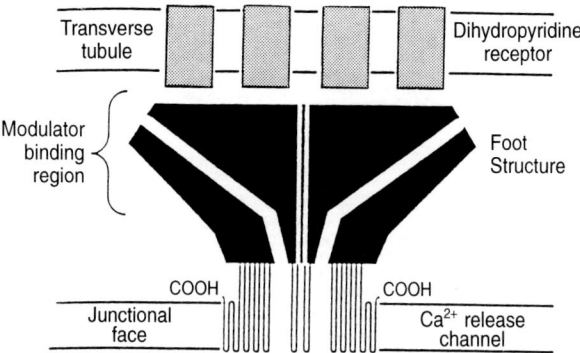

Fig. 3.4. Diagrammatic representation of the ryanodine receptor and its relation to the T-tubular tetrad, the dihydropyridine receptor. The bulk of the ryanodine receptor makes up the 'foot protein' lying in the junctional space which is a homotetramer with radial channels. Reproduced from MacLennan (1990).

detailed structural analysis of the isolated particles reveals a central hole of about 2 nm diameter with four more holes arranged symmetrically around it. The particles representing the 'feet' project about 12 nm from the outer face of the membrane whereas on the luminal side of the membrane there is an aggregation of calsequestrin (Fig. 3.4). The receptor particles are arranged in rows aligned at a constant distance. Close to them on the face of the membrane of the transverse tubule are groups of four particles that have been identified as dihydropyridine (DHP)receptors (for a review, see Campbell *et al.*, 1988). Dihydropyridines block the so-called L-type calcium channels that are responsible for the calcium inward currents of skeletal and cardiac muscle. Binding studies suggest that the ryanodine receptors are twice as abundant as the DHP receptors. The groups of four DHP receptors on the transverse tubule membrane, the tubular tetrads, with the associated membrane phospholipid correspond to the L-type voltage sensitive calcium channels.

Thus it would appear that a calcium channel exists in the transverse tubule that permits the entry of calcium into the cell and arranged close to it is the calcium release channel of the terminal cisternae. It is not known whether these two channels can act independently but there is evidence that in the cell the two systems are coupled together. The nature of the interaction between the two systems is obviously not a simple one to one arrangement as the ryanodine receptors are more abundant. Nevertheless, an attractive hypothesis is that the DHP receptor acts as a voltage sensor that detects the wave of depolarisation moving along the transverse tubule. In some way not understood, possibly by conformational change or liberation of a second messenger, this leads to changes in the ryanodine receptor causing the calcium release channel to open. No evidence of direct binding between the two receptors, however, has so far been obtained either from electron microscopy studies of the triads or from studies on the isolated receptors.

There have been a number of suggestions as to the nature of the signal that leads to the opening of the calcium channels associated with the ryanodine receptor in the SR membrane. Several investigators have demonstrated that the probability that the calcium release channels in the SR are open rises on increasing the Ca^{2+} concentration on the sarcoplasmic side. This explains the phenomenon of calcium-induced Ca^{2+} release (CICR) which was first observed by Ford & Podolsky (1970) in skinned muscle fibres. The experimental evidence indicates that CICR can occur in skeletal muscle. There is, however, considerable doubt whether it is an essential step in excitation–contraction coupling because contraction of frog muscle can occur in calcium-free saline. CICR may be much more important in cardiac muscle where calcium in the perfusate is an essential requirement for contraction of the isolated heart.

The established role of inositol 1,4,5-trisphosphate ($InsP_3$) in triggering the release of Ca^{2+} to initiate a wide variety of cellular responses has raised the question of whether it has a similar function in excitation–contraction coupling in muscle. The enzymes for inositol triphosphate release and degradation are present in skeletal muscle but at rather low levels. For example $InsP_3$-5 phosphatase activity is too low by several orders of magnitude to account for the speed of the termination of Ca^{2+} release during the relaxation of skeletal muscle. *In situ* release of $InsP_3$ by photolysis of caged $InsP_3$ gave a contractile response likewise very much slower than the normal response (Walker *et al.*, 1987). In contrast on photolytic release of $InsP_3$ in smooth muscle the contractile response was very similar to that obtained by the external application of adrenaline. This observation and the fact that the level of $InsP_3$-5 phosphatase is about 30 times higher in smooth than in skeletal muscle suggests that the $InsP_3$ pathway for Ca^{2+} release is important for the former tissue.

Experiments with isolated SR and skinned fibres have indicated that calcium release is sensitive to H^+ concentration.

This could be explained on the assumption that the channels open in response to a proton gradient across the membrane. It is unlikely that the pH changes measured on a single twitch, which usually represent a value averaged over the whole cell, would be adequate to establish the gradient required. Transient localised pH changes in the region of the triad associated with membrane depolarisation, however, could be higher and adequate to bring about the required effects. The significance of proton gradients for calcium release by the SR has yet to be assessed.

The mechanisms of release described above relate to the calcium release channels identified as associated with the ryanodine receptor. An additional mechanism for the release of calcium into the sarcoplasm could involve the reversal of the pump. The ability to reverse the SR calcium pump has been well demonstrated by *in-vitro* studies but it is unlikely that this plays a significant part in the release of calcium on contraction. Even under optimal conditions for reversal of the pump, which do not apply in intact muscle, the efflux of calcium is 10^2 to 10^4 times slower than the calculated rate of release during activation (Martonosi, 1986).

Calcium-binding proteins of muscle

To initiate the events that lead to contraction when the Ca^{2+} concentration rises above the resting level the ions must interact specifically with a component of the system, usually protein, that consequently becomes modified in some manner. This change, probably conformational, initiates effects in other components of the regulatory system that activate the Mg ATPase of the actomyosin system. The calcium ion has certain properties (Table 3.1) that enable it, when present at concentrations in the micromolar range, to bind specifically with high affinity to sites on a special group of proteins, the calcium-binding proteins. The fact that calcium ions can do this, even in the presence of magnesium and alkali metal ions in the

Table 3.1. *Comparison of the properties of calcium and magnesium ions*

	Ca^{2+}	Mg^{2+}
Intracellular concentration	10^{-7}–10^{-5}	10^{-3}
Enzymic function	Trigger	Substrate complex
Ion size (nm)	0.099	0.065
Coordination number	8 (6–10)	6
Bond distances and angles	Flexible	Restricted
Ability to cross link between organic groups	Good	Poor

millimolar and decimolar ranges respectively, is a reflection of the flexibility in coordination number, bond distances and angles accommodated by the ion when acting as a ligand.

α-Helix-calcium-binding loop-α-helix proteins

All proteins binding calcium with high affinity possess one or more characteristic domains to which the ion binds. These domains which were first identified by Kretzinger & Nockolds (1973) in carp parvalbumin consist of a peptide sequence of about 30 amino acid residues. They are very homologous in different proteins and consist of a basic structure of α-helix-calcium-binding loop-α-helix (HLH). The calcium-binding loop consists of about 12 contiguous amino acid residues which provide the oxygen ligands to the metal ion (Fig. 3.5). Usually there are seven ligands but not all have to be from the protein. Many of the binding sites have one or more water molecules in the coordination sphere of the calcium ion. In those coordination cages with seven ligands the oxygen atoms are located approximately at the seven vertices of a pentagonal bipyramid. The refined crystal structures for calcium-binding proteins now available show that there are seven oxygen atoms all at an average distance of 0.24 nm from the calcium ion. Five of them in an approximate pentagonal arrangement lie close to a common plane that includes the calcium ion. The vector joining the other two oxygen atoms is approximately at right angles to the pentagonal plane and passes close to the calcium ion. The HLH family exhibits a characteristic deviation of one of the five ligands from the pentagonal plane. This deviation may be important in enabling the site to bind magnesium as well as calcium as occurs in some domains, for example sites III and IV in troponin C (Strynadka & James, 1989).

Three proteins possessing HLH domains with high affinity for calcium and that play important parts in the regulation of contraction have so far been identified in muscle (Table 3.2). The first of these, calmodulin, is not unique to muscle but widely distributed in all tissues as an essential component of a variety of calcium-dependent processes. Its function in regulation is as an essential component in the myosin light chain kinase enzyme system, which is of particular importance in the regulation of smooth muscle. For this reason the concentration is higher in smooth muscle than in any other muscle. Troponin C is the highly specialised calcium-binding protein of the troponin complex, the immediate target for calcium when the level rises in striated muscle. So far significant amounts have

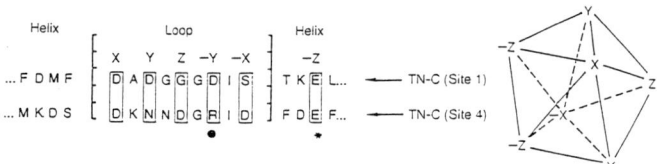

Fig. 3.5. Sequences of calcium-binding domains in troponin C. The coordinating residues are indicated as X, Y, Z,-X,-Y and -Z in the pentagonal bipyramid. With the exception of position -Y, where it is main chain oxygen, all coordinating groups are side chains. Both oxygens of the side chain in position -Z are involved in coordinating metal. Reproduced from da Silva & Reinach (1991).

Table 3.2. *Calcium-binding proteins of special significance for the regulation of contractile activity in muscle*

Protein	Molecular weight	Fast skeletal (Concentration μM)	Smooth	Ca^{2+} binding sites per mol	Affinity
Calmodulin	16 700	3	15–30	4	High
Troponin C	18 000	70	—	4(3 cardiac)	High
Parvalbumin	9–13 000	600	—	2	High
Regulatory light chain	18–20 000	400	100–150	1	Low

only been isolated from striated muscle. There is no clear evidence that troponin C is present in vertebrate smooth muscle, but if it is the amounts are very low. It is of interest that small amounts of the mRNA for the slow skeletal muscle isoform of troponin C have been detected in certain human fibroblasts.

When the amino acid sequences of the myosin light chains are examined a rudimentary calcium-binding site of the HLH type can be detected. The light chains are therefore considered to have evolved from the same primitive gene as the other calcium-binding proteins but when separated from myosin their calcium-binding constants are quite low. Nevertheless, the regulatory light chains of myosin, that is, the P light chain of vertebrate myosin, and the homologous EDTA light chain of molluscan adductor myosin, are included in Table 5.1. The isolated light chains from smooth and vertebrate myosins do not bind calcium with high affinity, but in the intact head of molluscan adductor myosin they interact with the heavy chain to form a novel calcium-binding site of high affinity (Chapter 5). In this way they are able to confer calcium sensitivity on the actin-activated ATPase of this system (Lehman *et al.*, 1972). There is evidence that under some conditions phosphorylation of the P light chain changes the calcium sensitivity of the vertebrate skeletal muscle (Sweeney & Stull, 1990).

The HLH domains of the calcium-binding proteins in all cases bind calcium with dissociation constants in the range 10^{-5} to 10^{-9} M. At some HLH domains magnesium is also bound (Fig. 3.6), probably as a consequence of a slight change in the positions of the ligands (Strynadka & James, 1989). In these cases the binding constant for magnesium is lower than that for calcium but high enough to ensure that at the normal resting intracellular concentrations of the two cations, magnesium occupies the site. When the Ca^{2+} concentration rises it will displace magnesium from the site but the process will be relatively slow because the off-rate for the latter ion is slow. For this reason such sites will not be effective in trigger-like responses to a rise in calcium concentration but will have a buffering role. In this respect it is significant that contraction is initiated in striated muscle by calcium binding to the calcium-

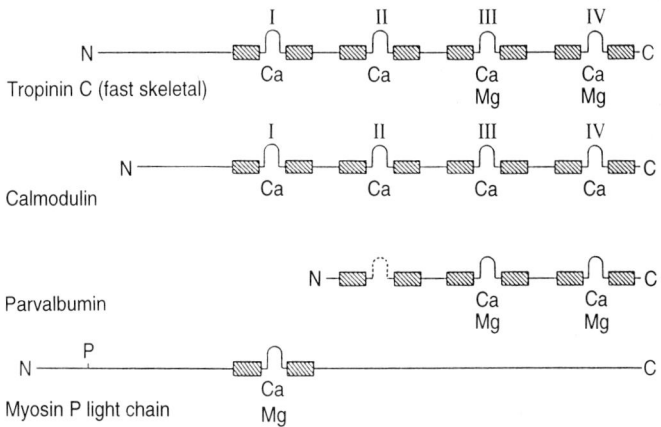

Fig. 3.6. Schematic representation of the polypeptide chains of the HLH proteins of muscle indicating the types and distribution of the calcium-binding domains. Reproduced from Perry (1994).

specific sites I and II on troponin C (Fig. 3.6). The presence of two types of site would appear to have significance for the contraction–relaxation cycle of striated muscle for such a distinction does not appear to exist between the four calcium-binding sites of calmodulin. It seems likely that troponin C evolved from a calmodulin-like precursor by mutations that changed the character of sites III and IV so that in resting muscle these sites are filled with magnesium. This confers a conformation change on the C-terminal half of the molecule and exposes hydrophobic groups (see later). These sites equal in number the sites that trigger contraction when filled with calcium. They would be expected to play a role in the relaxation process for they possess a slightly higher binding constant for calcium than sites I and II. Therefore as the Ca^{2+} concentration remains elevated while contraction progresses, the magnesium will be displaced by calcium. If the calcium flux from the SR is limiting the cation will move from sites I and II to sites III and IV because of the higher affinity of the latter sites.

The recent determination of the high resolution structure of fast skeletal muscle troponin C (Herzberg & James, 1988; Satyshur *et al.*, 1988) has given insight into the molecular changes that may occur in the protein when muscle is stimulated. The crystals studied contained calcium in sites III and IV but not in sites I and II (Fig. 3.7). It is reasonable to conclude that the conformation of troponin C determined from these crystals corresponds to that in resting muscle where the C-terminal sites III and IV are also occupied by cation, although in this case magnesium. On the basis of the high degree of sequence homology and the extensive structural similarity of the calcium-filled domains it has been proposed (Herzberg *et al.*, 1986) that the N-terminal calcium-specific domains take up a conformation similar to that exhibited by the two cation-filled C-terminal sites. The latter occupy a rather open structure with hydrophobic regions exposed to the solvent, a similar arrangement to that shown in Fig. 3.8(*b*). In

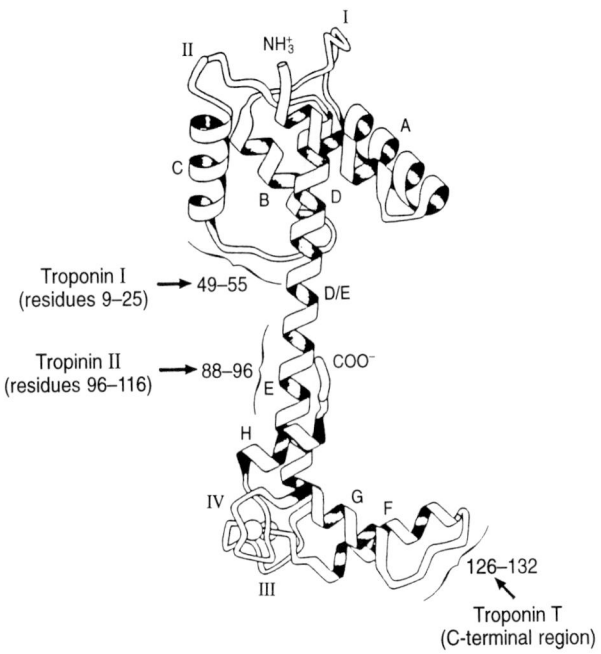

Fig. 3.7. Ribbon representation of the polypeptide chain of turkey skeletal muscle troponin C. The upper domain is the calcium regulatory domain (N-domain); the lower domain is the high affinity Ca^{2+}/Mg^{2+}-binding domain (C-domain). The eight helices involved in the four supersecondary helix-loop-helix motifs are labelled A–H sequentially. There is an additional helix at the beginning of the N-domain sequence. The segments of the polypeptide chain forming the calcium-binding loops are labelled I to IV. The numbers of the residues of troponin C and I corresponding to the presumed interaction sites are those determined for the fast skeletal muscle isoforms of the proteins obtained from the rabbit. Reproduced from Perry (1994), modified from Herzberg & James (1988).

contrast, the two unoccupied N-terminal domains are arranged in a compact structure with hydrophobic surfaces unavailable to the solvent (Fig. 3.8(a)). When the sites fill with calcium reorientation of the helices A/B and C/D occurs producing a more open structure in which hydrophobic residues are exposed to solvent, as is the situation in the C-terminal part of the molecule. This change is the consequence of residues comprising the peptide-linking helices B and C moving by up to 1.4 nm. According to the model the N-terminal helix N does not change in orientation relative to helix D and helix pairs A/D and B/C also retain their relative orientations (Fig. 3.8(b)). The conformational changes described above are probably not unique to troponin C in view of the homology of the calcium-binding domains in the HLH proteins.

Support for the hypothesis that relative movement between

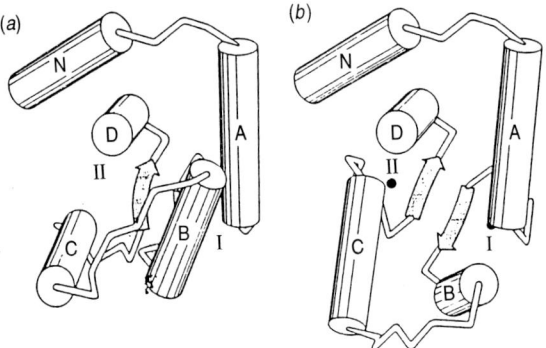

Fig. 3.8. Diagrammatic representation of the proposed Ca^{2+}-induced change in the N-terminal domain of troponin C. In this model helices N, A and D retain their relative positions. Helices B and C and the linker peptide move up to 1.4 nm when Ca^{2+} binds. The relative positions of helices B and C also remain constant. (a) Ca^{2+}-free conformation of the N-terminal domain of troponin C. (b) Proposed Ca^{2+}-bound form of this domain. Reproduced from Strynadka & James (1989).

calcium-binding sites I and II is involved in the function of troponin C has been obtained using site-directed mutagenesis of the troponin C gene (Grabarek *et al.*, 1990). By replacing glutamine residues at positions 48 and 82 with cysteines a mutant was obtained which when oxidised formed a disulphide bond linking the cysteine residues and thereby locking the two N-terminal calcium-binding sites in one position. The oxidised mutant, although it can bind to troponin I, did not restore calcium sensitivity to troponin C-deficient myofibrils whereas the wild type troponin C did. Thus a common feature of calcium binding by these proteins is the exposure of hydrophobic regions which enable interactions to occur with other components of the system of which they form a functional part (for a review, see Grabarek *et al.*, 1992).

Parvalbumins

The parvalbumins are the prototype calcium-binding proteins. They were discovered as the major soluble proteins in the white muscle of fish (accounting for about 0.7% of the weight of carp muscle, for example) long before their function was understood. It was originally thought that the parvalbumins were confined to the muscles of fish, amphibia and certain reptiles and that their role was in some way related to the particular way of life of these creatures. Parvalbumin has now been shown to be present in the fast white muscle of many mammals including humans although usually in much smaller amounts than is the case with fish. In fish it occurs in a number of isoforms whereas only a single isoform has been detected in the muscles of mammals. The amounts are higher in the type 2 fibres of small mammals but very low in muscles of large animals even if they consist predominantly of type 2 fibres. Parvalbumin cannot be detected in cardiac or smooth muscle cells but surprisingly a subpopulation of brain cells do stain with antibody to the protein (Heizmann, 1984).

Unlike troponin C which is built into the contractile structure of the muscle, parvalbumin is not involved in the initiation of contraction. It is dissolved in the sarcoplasm surrounding the myofibril and in some muscles at higher concentrations than troponin C. The cation-binding properties of the two HLH domains are very similar to those of sites III and IV of troponin C. As is the case with the latter sites in resting muscle, those in parvalbumin are occupied with magnesium which in the later stages of the contraction–relaxation cycle is displaced by calcium. Thus it acts as calcium buffer and would, if the concentration of parvalbumin relative to troponin C is high enough, effectively speed up relaxation. It has been concluded from computer simulation studies that immediately after a pulse of Ca^{2+} corresponding to that occurring on contraction that the regulatory sites I and II of troponin C would be 97% saturated but very little calcium would be bound to parvalbumin (Gillis & Gerday, 1977). After 200 ms, which is comparable to the duration of a twitch in frog sartorius muscle, the magnesium on the parvalbumin would be replaced by calcium. This study suggests that in muscles with parvalbumin concentrations at least five times greater than that of troponin C, the parvalbumin acts as a soluble relaxing factor. Binding of calcium to parvalbumin will occur towards the end of contraction when the Ca^{2+} concentration in the sarcoplasm begins to fall and hence lead to a speeding up of relaxation. Such an explanation of its role is supported by the correlation that exists between relaxation speed and parvalbumin content in mammalian skeletal muscles, although the amounts present tend to be much lower than in fish muscle.

Calsequestrin

The major protein component of the lumen of the terminal cisternae is calsequestrin, a calcium-binding protein of much lower affinity than the HLH group of proteins. The dissociation constant is higher, 10^{-3}, but the capacity of this protein for calcium is much greater than the high affinity proteins. The calsequestrin of rabbit fast twitch muscle with a molecular mass of 42 kDa binds up to 40–50 moles of calcium per mole. Like many muscle proteins it exists in several isoforms that are characteristic for the muscle type. The calcium-binding properties of calsequestrin are a consequence of its high content of acidic amino acids; for example in the rabbit fast skeletal muscle isoform only 15% of the aspartic and glutamic acid residues are amidated with the result that the protein has a strong net negative charge (Fliegel *et al.*, 1987). Its function is clearly to act as a calcium buffer in the SR with the property of readily releasing the cation when the calcium channels in the SR are opened. There is evidence that in the absence of calcium calsequestrin binds to proteins of the junctional SR (Mitchell *et al.*, 1988). This suggests that the protein is anchored close to the calcium pump in order to sequestrate the calcium as it is removed from the sarcoplasm. When calcium is bound the calsequestrin no longer interacts with the junctional proteins and disperses into the lumen of the cisternae. Although the acidic residues are clustered the amino acid sequence of the protein does not suggest that the binding sites are of the HLH type. Considerable conformational change accompanies the binding of calcium but unlike the situation in the HLH group of proteins it is accompanied by a loss in hydrophobicity.

Kinetics of calcium release

To understand the precise relationship between sarcoplasmic Ca^{2+} concentration and force generation in muscle it is necessary to measure the ion concentration during the contraction–relaxation cycle. There are two main types of reporter molecules that can be introduced into muscle to produce signals from which the intracellular Ca^{2+} concentration can be followed. The first of these are the chemiluminescent proteins which on binding Ca^{2+} emit light, the intensity of which is a measure of the free Ca^{2+} concentration. Aequorin, a protein of molecular mass of about 20 kDa prepared from the jelly fish, *Aequorea forskalea*, has been widely used to study the Ca^{2+} transients in skeletal muscle, particularly in the giant fibres of barnacle muscle (Ashley *et al.*, 1991). The size of these cells facilitates the injection of the relatively large amounts of aequorin required as each molecule emits at most one photon. The other group of probes includes fluorescent molecules such as fura-2, indo-1 and fluo-3, into which a selective calcium-binding site modelled on the calcium chelator EGTA has been incorporated (Tsien, 1988). The large spectral changes that occur on binding Ca^{2+} by these compounds can be used to measure the concentration of the ion. The advantages of the fluorescent probes are their low molecular mass (less than 1 kDa), their ability to be re-excited 100 to 10 000 times before being bleached and the ease with which they can be introduced into a cell. On incubation lipophyllic derivatives such as esters readily pass into the cell without damaging the integrity of the membrane.

The use of such probes indicates that the Ca^{2+} concentration in resting skeletal muscle is 0.07–0.1 μM, rising to 2–10 μM during contraction. Taking average values for the resting and contracted states this rise in concentration requires at least 6 μmoles of Ca^{2+} per kilogram of muscle to move from the SR to the sarcoplasm during contraction. The actual flux must be much greater for the two regulatory N-terminal calcium-binding sites of troponin C which must presumably both be filled before contraction can take place. In addition, the small amounts of calmodulin (four calcium-binding sites per molecule) present in skeletal muscle will also be saturated with calcium. On the basis of the concentrations of the calcium-binding proteins given in Table 3.2 this means that approximately an additional 150 μmoles of Ca^{2+} per kilo of muscle must be transferred from the SR when contraction occurs. If as might be expected in the later stages of the cycle calcium replaces magnesium from the high affinity sites of troponin C another 140 μmoles will be bound to the latter protein. This can be provided from the SR stores and without loss from the low affinity sites on troponin C, because the SR present in a kilo of skeletal muscle can store $1000\,\mu M$ Ca^{2+} (Martonosi, 1986). The situation is very different, for example in muscles rich in parvalbumin, such as that of the carp, where the concentration of the latter protein may be five to ten times that of troponin C. In this tissue not enough calcium is available from the SR to saturate all the calcium-binding proteins present. The binding of calcium to the high affinity sites on the excess of parvalbumin present will therefore lead to its loss from the regulatory sites of troponin C and relaxation of the muscle.

Studies with the giant fibres of barnacle using aequorin (Ashley *et al.*, 1991) indicate that the relation between force developed and free calcium ion concentration is not simple (Fig. 3.9). On application of a single voltage-clamp impulse to a barnacle muscle fibre the free Ca^{2+} concentration rose rapidly but there was an appreciable delay in force development and the

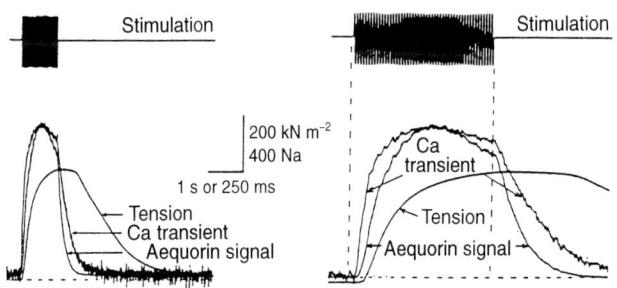

Fig. 3.9. Aequorin light responses, free Ca^{2+} concentration and tension during electrical stimulation of an intact barnacle muscle fibre. Free calcium transient obtained from the aequorin signal and its first derivative. Reproduced from Griffiths *et al.* (1990).

Ca^{2+} concentration passed its peak before the force developed reached its maximum. These observations suggest that the Ca^{2+} release by the SR is extremely rapid. As might be expected the observations indicate that the rate limiting processes for force development during contraction lie in the chain of events from the binding of calcium to the target proteins to the modulation of the interaction of actin and myosin that results in a high rate of ATP hydrolysis. There is a similar lack of close correlation between force developed and Ca^{2+} concentration during the relaxation phase because the free Ca^{2+} concentration falls more rapidly than the force. This implies that the reversal of the process that leads to the modulation of the interaction of actin and myosin as a consequence of the loss of Ca^{2+} from the troponin C occurs at a slower rate than it is pumped back into the SR. If the muscle contains parvalbumin, the loss of calcium from this protein would also be a factor determining the Ca^{2+} transients in relaxation. In barnacle muscle in which much of the work with aequorin has been carried out this is not a significant factor as the parvalbumin concentration is very low (less than $6\,\mu M$).

4

I filament regulation

In the presence of ATP at the appropriate ionic conditions
model systems composed of purified actin and myosin isolated
from striated muscle invariably undergo contraction accompanied by a high rate of ATPase activity. Soon after the original
development of such model systems (Szent-Gyorgyi, 1945) it
was shown that relaxation could be induced or contraction
prevented by inhibitors of the myosin ATPase. These observations and the correlation of contraction with high enzymic
activity suggested that stimulation of muscle resulted in the
lifting of the inhibition of the actomyosin MgATPase that
existed in the resting state. It was not, however, until about 20
years later that the components involved in regulation of the
myofibrillar ATPase in intact muscle were precisely identified
(for a review, see Perry, 1994).

Marsh (1952) was the first to demonstrate that a sarcoplasmic extract contained a factor or factors that were able to
prevent a crude myofibrillar fraction of rabbit muscle from
responding to the addition of ATP by contraction. Subsequently the particulate component of the sarcoplasm that we now
identify as the SR, and which has the ability to concentrate and
release calcium, was shown to be an important component of
the system (Kumagai *et al.*, 1955). Somewhat earlier studies on
the effects of the microinjection of cations into living muscle
fibres had indicated that calcium was particularly effective in
causing contraction. Up to that time there was, however, no
direct evidence that calcium was the trigger for the contractile
response or that it was required for the MgATPase activity of
the purified actomyosin systems. It was reported by Perry &

Grey (1956*a*, *b*) that the MgATPase activity of crude preparations of actomyosin, euphemistically called 'natural actomyosin', was inhibited by low concentrations of EDTA and
EGTA (then known as glycolkomplexon). Actomyosin prepared from purified preparations of the two proteins did not
exhibit this property. This implied that the MgATPase activity
of crude actomyosin preparations was calcium sensitive, an
observation that was further substantiated in detail by Weber
(1959). Ebashi (1963) later showed that the calcium sensitivity
of the MgATPase of 'natural actomyosin' resulted from the
presence of the troponin complex and tropomyosin. Thus it
became apparent that the trigger for muscle activity was the
calcium ion and the mechanism involved the SR and proteins
located in the myofibril.

Regulatory proteins of the myofibril

Tropomyosin

When tropomyosin was discovered and characterised in striated muscle by Bailey (1946) its physical properties suggested
that it had a structural function in muscle. It was not until over a
quarter of a century later that it became apparent that
tropomyosin also has a role in the regulatory process in striated
muscle. Unlike the troponin complex, which appears to be
unique to striated muscle and has evolved to accommodate the
special contractile properties of that tissue, tropomyosin is
much more widely distributed. It is present in smooth muscle
and also in non-muscle tissues but in smaller amounts than in
skeletal muscle where it accounts for up to about 3% of the total
protein. Due to its special affinity for actin it is probably
associated with all contractile systems involving actomyosin
and cytoskeletal systems in which actin is implicated. It is not
possible to decide whether its role in striated muscle is unique to
that tissue or simply an adaptation of its property of interacting

with actin which can be utilised for other intracellular functions.

Tropomyosin is an asymmetric molecule of dimensions 40×2 nm and molecular weight approximately 66 000. Its molecular structure is unusual in that it is a dimer consisting of two α-helical polypeptide chains coiled around each other in parallel to form the so-called coiled-coil structure. As a consequence the molecule is $>95\%$ α-helix. The polypeptide chains of the subunits exhibit a characteristic heptapeptide repeat with non-polar residues in positions 1 and 4 that correspond to the points of contact between the two α-helices (Smillie, 1979). The coiled-coil structure is stabilised by the juxtaposition of polar and ionic residues in other positions of the heptapeptide. By analysis of the X-ray diffraction patterns of living muscle and image reconstruction analysis of electron micrographs of the reconstituted thin filaments containing actin and the regulatory proteins, the location of tropomyosin in the I filament has been determined. It lies as a linear polymer, one molecule thick, in each of the two grooves of the actin double helix (Fig. 4.1). In effect a tropomyosin molecule lies alongside seven actin

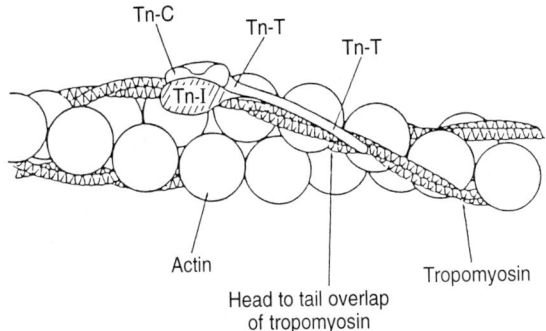

Fig. 4.1. Schematic representation of the thin filament of vertebrate striated muscle indicating suggested locations of the components of the troponin complex. Reproduced after Heeley *et al.* (1987).

subunits that are present in every 38.5 nm of each of the two threads of actin monomers that make up the F-actin filament. The molecular length of tropomyosin is 40 nm but the axial spacing of troponin with which the tropomyosin is associated is 38.5 nm, that is, very similar to the half pitch of the actin filament for which values of 35–40 nm have been reported (Haselgrove, 1983). The reason why the effective length of tropomyosin in the filament is less than its molecular length is probably a consequence of the way in which the protein is arranged in the groove and because the molecule polymerises in the I filament with a short head-to-tail overlap involving 8–11 residues of the N and C termini. In consequence, the effective molecular length is similar to the filament repeat and the molar ratio of actin to tropomyosin is 7:1.

On stimulation changes occur in the X-ray diffraction pattern of muscle which are interpreted to indicate that the tropomyosin has moved from the edge into the centre of the groove, in all a distance of 1.0–1.5 nm. This is a very early change in protein distribution, occurring 17 ms after stimulation. The myosin head attaches to actin at 28 ms and tension develops at 40 ms. The evidence for movement of tropomyosin as an early event in the excitation process has led to the proposal of the steric mechanism for the regulation of contraction (Haselgrove, 1972; Huxley, 1972). This postulates that in resting muscle tropomyosin blocks the site on actin with which the myosin head must interact to activate the MgATPase. On stimulation it is presumed that tropomyosin moves from the blocking position to enable the myosin head to interact with actin and the cross-bridge cycle to proceed. Although this hypothesis is attractive in its simplicity, full acceptance of it in its original form requires a clear demonstration that: (1) the position of tropomyosin in resting muscle is that required to block the site on actin with which myosin binds to activate the ATPase, and (2) that on stimulation the tropomyosin moves to leave the site available for interaction.

With the three-dimensional structure of actin now available at 0.28 nm resolution and the information that is currently accumulating regarding the amino acid residues involved in the interaction with myosin, this problem should soon be resolved. There is convincing evidence for the movement of tropomyosin in the I filament in response to calcium activation, both from X-ray diffraction and electron microscopic image reconstruction studies of vertebrate muscle. Further recent image reconstruction investigations with preparations from *Limulus* muscle (Lehman *et al.*, 1994) clearly demonstrate movement of the tropomyosin, which is compatible with the steric hypothesis, when thin filaments are activated by calcium. It should be pointed out, however, that there is some experimental evidence that argues against the steric mechanism, at least in its simplest form. The hypothesis would imply that the binding constant between actin and myosin would be much larger in the activated system, that is, in the presence of calcium, than in the relaxed state. Kinetic studies suggest that the difference in the binding constants between the two states determined using regulated *in-vitro* systems of actin and subfragment 1 or heavy meromyosin are much less than would be expected (Chalovitch *et al.*, 1981). It is of interest that similar changes in the position of tropomyosin in the thin filament, as indicated by the X-ray diffraction pattern (Lowy & Vibert, 1972), occur in smooth muscle. This tissue does not contain troponin and in the activation process occurs via the A filament. It must be concluded that if tropomyosin has a regulatory role in smooth muscle it is different from that proposed for striated muscle where its function would appear to be mediated through the troponin complex.

As is the case with most of the other myofibrillar proteins tropomyosin occurs in a number of isoforms, proteins which are clearly tropomyosin in general properties and molecular structure but differ slightly in amino acid sequence. In skeletal muscle α and β forms, each under different genetic control, are expressed. Both genes can give rise to additional isoforms by RNA splicing with the result that in skeletal muscle three or four isoforms can be sometimes detected. Whereas with some myofibrillar proteins such as troponin I and C a single isoform is present in a given cell type, both α and β tropomyosins are present in fast and slow skeletal muscle cells. The isoform composition is, however, specific for the muscle cell type in so far as the proportion of α-tropomyosin is higher in fast than in slow striated muscle cells. Modifications in the isoform composition accompany changes in the speed of muscle such as occur during development and after cross innervation (Heeley *et al.*, 1985). The association of specific combinations of isoforms with muscle fibres with different speeds of contraction and relaxation implies that the isoform composition may have a role in determining the nature of the physiological response of the fibre. A consequence of the fact that more than one isoform of tropomyosin subunit is present in a given muscle fibre is the probability that in the cell the tropomyosin molecule may exist as a heterodimer and in this way subtly modulate the properties of the dimer molecule. Further modification of the properties of tropomyosin may occur by phosphorylation of a penultimate residue, serine 283 in rabbit α-tropomyosin (Chapter 5).

Troponin complex

When Ebashi (1963) first isolated the protein fraction present in preparations of 'natural actomyosin' that conferred calcium sensitivity on its MgATPase he named it 'tropomyosin-like protein' because of its properties. Later he showed that this fraction consisted principally of tropomyosin and a new protein he called troponin. The latter protein complex, which forms a viscous complex with tropomyosin, is essential for conferring calcium sensitivity to the actomyosin MgATPase. Troponin is a complex of one molecule each of three monomeric proteins that have distinct but coordinated roles in the regulatory process.

Table 4.1. *Summary of the properties of the components of the troponin complex*

	Troponin C	Troponin I	Troponin T
Molecular weight	18 000	21 000–23 000	30 000–35 000
Role	Calcium binding	Inhibitory action	?
Interactions	Troponin I Troponin T	Troponin C Actin Troponin T (?)	Tropomyosin Troponin C Troponin I (?)
Phosphorylation sites	No	Yes	Yes

These are troponin C, troponin I and troponin T (Table 4.1). A study of the staining pattern of the I filament with antibodies to the troponin components by electron microscopy and X-ray diffraction indicates that the complex is distributed with a periodicity of 38.5 nm, that is, very similar to that of the F-actin filament, corresponding to half the pitch of the actin helix. As yet detailed three-dimensional structural data on the troponin complex is not available as troponin C is the only component so far to be crystallised and for which the structure at high resolution has been determined (Chapter 3). The present views on the arrangement of the regulatory proteins in the I filament of skeletal muscle are summarised in Fig. 4.1. Despite uncertainty as to the precise spatial interrelationships between the regulatory proteins of the I filament, the stoichiometry of seven actin monomers:one tropomyosin molecule:one molecule of each of the three troponin components is well established.

Troponin C

The binding of calcium to troponin C, the target protein of the complex (Chapter 3), initiates a complex series of events involving the proteins of the I filament that transform actin from its resting role into an activator of the myosin MgATPase. Significant amounts of troponin C appear to be confined to striated muscle where it represents from 0.5 to 0.7% of the total muscle protein, the highest amounts being present in fast twitch skeletal muscle.

As yet only two isoforms of troponin C, which are under different genetic control, have been isolated from muscle. These are the fast and slow twitch skeletal muscle forms. The latter is identical with the cardiac form and although it contains four HLH domains, that corresponding to calcium-binding site I of the fast muscle form is defective. It has the X and Y aspartic positions in the binding loop (Fig. 3.5) replaced by leucine and alanine which renders it relatively ineffective as a high-affinity calcium-binding site (Fig. 4.2). The consequence of this is that in slow skeletal and cardiac muscles the binding of one rather than two calcium ions at the N-terminal domain of troponin C is adequate to trigger the cross-bridge cycle.

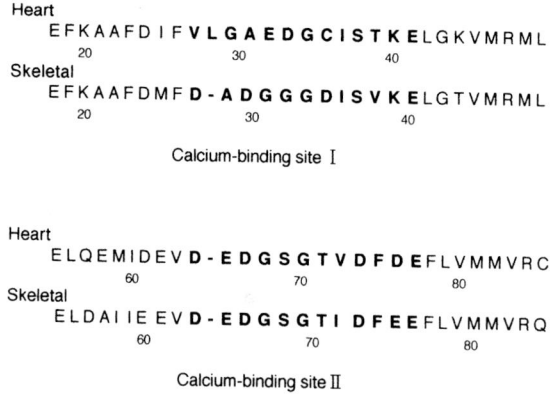

Fig. 4.2. Amino acid sequences of calcium binding sites I and II of rabbit skeletal and cardiac isoforms of troponin C. Residues in the calcium-binding loop are indicated by heavy type. Numbers indicate residue positions.

In addition to its calcium-binding properties troponin C also forms complexes with troponin I and T which are of functional significance (see later).

Troponin I

When isolated free from the other components of the complex troponin I has the property of inhibiting the actin-activated MgATPase of actomyosin. Unlike the inhibition obtained with the complete regulatory protein system the effect with troponin I alone is completely insensitive to calcium. Troponin I is a basic protein that occurs in a number of different isoforms with molecular masses in the range of 21 to 23 Kda. Usually the normal, mature striated muscle cell contains a single isoform

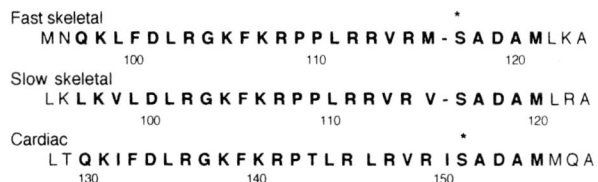

Fig. 4.3. Regions of the amino acid sequences of the isoforms of troponin I present in the striated muscles of the rabbit exhibiting close homology. Residues in these regions indicated by heavy type have common functional importance in all the isoforms (Figs. 4.4 and 5.2). Asterisks indicate serine residues that have been demonstrated to be phosphorylated *in vivo*. Presumably hydroxylated amino acids in homologous positions to the known phosphorylation sites are also phosphorylated. Numbers indicate residue positions.

that is specific for the cell type. The slow skeletal, fast skeletal and cardiac muscle isoforms of troponin I are each under individual gene control. As is the case with other myofibrillar proteins, procedures that produce changes in the muscle speed lead to the expression of the isoform appropriate for the acquired properties of the muscle. At the same time expression of the gene responsible for the isoform originally present is suppressed. The sequences of the isoform of troponin I are slightly less well conserved than those of troponin C but in the case of the rabbit are about 60% identical. Two regions of the sequence of functional significance, corresponding to residues 5–27 and 96–121 in the rabbit fast skeletal isoform, are strongly conserved (Fig. 4.3).

The inhibitory activity of troponin I is much increased by tropomyosin and the data obtained from actomyosin systems reconstituted *in vitro* suggest that in the presence of the latter protein one molecule of troponin I can prevent the activating action of more than one monomer of actin on the MgATPase of myosin. Usually the experimental values obtained *in vitro* with the isolated proteins indicate that one troponin I molecule can prevent the activating effect of three to four actin monomers, although the structural evidence suggests that in the intact myofibril the inhibitory activity of one troponin molecule regulates seven actin monomers.

Interaction of troponin I with actin and troponin C

Troponin C forms a complex with troponin I that is strengthened in the presence of calcium. Indeed, in the presence of the latter cation the complex is so strong that it is not dissociated by 8 M urea. The importance of complex formation for the regulatory process is indicated by the fact that when it occurs the inhibition of the actomyosin MgATPase by troponin I in the presence or absence of tropomyosin is relieved. The blocking of the inhibitory action by troponin C occurs both in the presence and absence of calcium. This suggests that even in the absence of this cation troponin C binds to isolated troponin I in a manner that prevents its inhibitory action. With the intact troponin complex the inhibitory action of troponin I is calcium sensitive, from which it follows that the presence of the third component of the complex, troponin T, modifies the interaction of troponin C with troponin I in such a way that the inhibitory action of the latter protein is controlled by calcium. Another consequence of the formation of the complex between troponin I and C is that the calcium affinity of troponin C is increased by a factor of ten.

In view of its key role in regulating the actomyosin Mg ATPase, knowledge about the mode of action of troponin I should provide insight into the mechanism of the transduction process. Investigation of this aspect is facilitated by the finding that the functional activity of the conserved regions of troponin I are in large part preserved in peptides corresponding to these regions obtained either by synthesis or by selective cleavage of the protein. For example, the peptide consisting of residues 96–116 of rabbit fast muscle troponin I obtained by cyanogen bromide cleavage of the intact molecule possesses inhibitory activity that is stimulated by tropomyosin (Syska *et al.*, 1976). It is significant that this peptide is derived from the region that is strongly conserved in all isoforms, residues 96–121 in the rabbit fast skeletal muscle troponin I. The fact that the inhibitory effect of such a small molecule can be extended by tropomyosin over a number of actin monomers must have important implications for its mode of action. Clearly in the presence of tropomyosin the peptide is capable of modifying the properties of actin monomers with which it is not closely associated.

By the application of proton nuclear magnetic resonance (NMR) it can be shown that the inhibitory peptide interacts with at least two regions of the actin molecule, both of which are exposed in the F-actin filament (Levine *et al.*, 1988). The N-terminal portion of the inhibitory peptide interacts with the region represented by residues 1–7 and the C-terminal portion with the region that contains residues 24 and 25 of actin (Fig. 4.4). Interaction with the C-terminal portion of the inhibitory peptide would appear to be of particular functional importance as it has been shown by studies with synthetic peptides that the inhibitory activity involves residues 105–114 in rabbit fast skeletal troponin I (Talbot & Hodges, 1981). There is evidence that the head of the myosin molecule, subfragment 1, can also interact with the regions of actin that are similar to those that bind to troponin I. This would indicate that a direct steric blocking of the myosin interaction site on actin by troponin I could explain the inhibition of the MgATPase. The explanation, however, may not be quite so simple because there is evidence that allosteric effects are associated with the interaction of these two proteins. For example, when troponin I or the inhibitory peptide binds to actin the interaction between the C-terminus of actin, which is not a binding site for troponin I, and the alkali light chain of myosin is inhibited (Grand *et al.*, 1983).

Peptides corresponding to residues 1–21 and 96–118 of rabbit fast muscle troponin I obtained by digestion with cyanogen

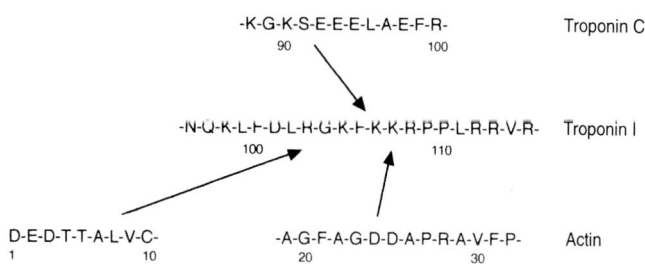

Fig. 4.4. Regions of the sequences of actin and troponin C shown by nuclear magnetic resonance (NMR) studies to interact with the inhibitory peptide of rabbit fast skeletal muscle troponin I (residues 96–116). Numbers indicate residue positions.

bromide can be shown by affinity chromatography and proton NMR to bind to troponin C. The interaction of troponin C with the two sites on troponin I can take place independently although NMR studies indicate that the points of contact between the two proteins are fairly close together. In both cases the interactions of the troponin I peptides with troponin C require calcium but higher concentrations are required for interaction with the peptide consisting of residues 1–21 (Dalgarno *et al.*, 1982). It follows that the interaction of troponin C with the N-terminal site of troponin I will be formed later during stimulation and broken earlier during relaxation than is the interaction with the inhibitory region. The function of the N-terminal interaction site on troponin I is not clear for it is the binding of troponin C at the inhibitory peptide region that neutralises the inhibitory action.

The sites on troponin C concerned in the interactions with troponin I are rather less well defined than the complementary sites on the latter protein. In addition to the calcium-binding domains there are other regions of the polypeptide sequence that are identical or strongly conserved in the different forms of troponin C and that are presumably of functional significance. These regions correspond to residues 49–55 between calcium binding domains I and II, residues 88–96 between domains II and III and residues 126–132 between domains III and IV of rabbit fast muscle troponin C (Fig. 3.7). A fragment obtained by cyanogen bromide cleavage of troponin C consisting of residues 44–77 binds to the N-terminal binding site on troponin I. This includes the invariant region of troponin C and the hydrophobic residues that become exposed when calcium is bound to sites I and II. It seems likely that this conformational change occurring at or very close to the interaction site is responsible in part at least for the great increase in affinity between troponins I and C when calcium is bound. Another cyanogen bromide peptide consisting of residues 88–137 forms a calcium-sensitive complex with troponin I and neutralises its inhibitory activity. All the evidence is that this peptide interacts with the inhibitory peptide region of troponin I and that the invariant residues 88–96 represent an important part of the interaction site on troponin C. These results are summarised in Fig. 3.7 in which the troponin C molecule is shown in the extended form as deduced from the crystal structure. The two points of attachment between the proteins are probably less widely separated than is illustrated in Fig. 3.7. There is evidence that troponin C may be in a more compact form in solution, and when complexed with troponin I, than is revealed by the crystal structure. Certainly proton NMR studies with troponin C spin-labelled at cysteine 98 indicate that both binding sites are close to this residue, that is, within 1.5 nm (Dalgarno *et al.*, 1982).

Recent studies on the skeletal muscle proteins with mutants and using cross-linking agents (Kobayashi *et al.*, 1994) indicate that interaction with the inhibitory peptide of troponin I involves both the N and C-terminal domains and the linker region of troponin C. In the complex the peptide sequences of the two proteins run antiparallel. NMR studies of the troponin I complex from cardiac muscle suggesting the N-terminus of troponin I interacts with C-terminus of troponin C imply a similar antiparallel arrangement (Krudy *et al.*, 1994).

It is clear from the properties of troponin I that it occupies a central role in the troponin–tropomyosin regulatory system of the I filament. Troponin I possesses a site that will bind either to actin, at a site which interacts with myosin to produce activation of the MgATPase, or to troponin C. These properties suggest a mechanism for controlling the cross-bridge cycle. In this context it is of interest that the proton NMR studies indicate that the hydrophobic residues in the N-terminal region of the inhibitory peptide are perturbed by troponin C whereas when interaction with actin occurs the arginine residues located mainly in the C-terminal region are affected. The latter observations correlate well with the observation that the inhibitory activity is restricted to a synthetic peptide consisting of residues

104–115 which contains four of the five arginine residues in the inhibitory peptide. In the light of this information it can be presumed that when the calcium level rises on stimulation troponin C binds to the inhibitory peptide region of troponin I displacing the actin bound at the C-terminal region. The interaction occurs in such a way that the site on actin now becomes available to interact with myosin and activate the MgATPase. Actin becomes available to activate the myosin MgATPase of fast skeletal muscle only when binding sites I and II of troponin C are filled with calcium. In resting muscle these sites do not contain calcium and the interaction of troponin C with troponin I is modified in such a manner that the latter protein binds to the myosin-activating site on actin with the result that the MgATPase is inhibited.

Troponin T

Troponin T, like troponin I, exists in three main forms that are under different gene control and are specific for the three types of striated muscle cell: fast, slow and cardiac. Consisting of a single polypeptide chain, troponin T is the largest of the components of the troponin complex with its isoforms possessing molecular masses in the range of 30 to 35 kDa. Troponin T is unusual among the troponin components in that the mRNAs, corresponding to the tissue-specific proteins, undergo alternative splicing procedures giving rise to a number of isoforms. It has been estimated that potentially 64 mRNAs can be generated in rat skeletal muscle from troponin T gene (Breitbart *et al.*, 1988). In vertebrate muscles usually three to five isoforms can be identified on one-dimensional electrophoresis of extracts of a given muscle type. Occasionally investigators using high resolution two-dimensional electrophoresis report many more.

In some muscle cells, such as the fast fibres of rabbit skeletal muscle, the proportions of the fast troponin T isoforms vary. Certain fast isoforms predominate when the skinned fibre possesses a steep Ca^{2+}/tension curve whereas others are associated with a less steep response (Schachat *et al.*, 1987). These observations may reflect the role of troponin T in modulating the effect of calcium on the troponin I–troponin C interaction and may provide further evidence of how slight sequence differences in the isoforms of the troponin components can modulate the physiological properties of the muscle. It is significant that those fibres exhibiting a steep Ca^{2+}/tension curve also contain a predominance of α-tropomyosin.

Troponin T resembles troponin I in that it is a basic protein with an isoelectric point in the region of pH 8. The evidence available suggests that it is an extended asymmetric molecule with relatively little ordered structure, although the region represented by residues 71–151 is about 80% α-helical. The extended structure is probably significant for its function, as are its properties of forming complexes with troponin C and tropomyosin. The interaction of troponin T with tropomyosin produces a viscous complex which is also obtained when the former protein is replaced with a peptide consisting of residues 71–160 obtained by cyanogen bromide digestion (Jackson *et al.*, 1975). An additional site for tropomyosin binding has been identified close to the C-terminus of troponin T (Pearlstone & Smillie, 1981). From a variety of studies it would appear that the asymmetric troponin T molecule is located lying along about one-third of the C-terminal end of tropomyosin molecule, the association being stabilised by interactions between the α-helix of troponin T and the coiled-coil of tropomyosin. The tropomyosin–troponin T interaction serves to fix the position of the troponin complex on the I filament and may enable conformational effects to be transmitted between the two proteins. Indeed, it has been suggested that conformational changes occurring in the troponin complex when calcium is bound are transmitted to tropomyosin by troponin T. In this way the movement of tropomyosin on contraction could be explained. Even if this is the case the troponin

T–tropomyosin interaction cannot explain the property of the latter protein in extending the inhibitory action of troponin I over a number of actin monomers because this occurs in the absence of troponin T.

The other important role of troponin T is to modulate the role of troponin I on the actomyosin interaction, that is, to make it calcium sensitive. The property of forming a calcium-dependent complex with troponin C and the correlation of certain isoform complements with the steepness of the calcium/tension curve suggest a specific role in the calcium triggered events. The complex with troponin C is calcium dependent but the interaction, unlike the troponin I–C complex, is not stable in high urea concentrations. Studies with specifically cleaved fragments of troponin T suggest that troponin C binds at the C-terminal end of the molecule and in the region of residues 150–170 (Fig. 4.5). It cannot be decided whether these are two different sites or merely represent parts of the same site in the folded molecule where the two regions are in close proximity. More detailed information on the troponin T binding site(s) on troponin C would clarify this situation (for a review of the interactions of troponin T, see Perry, 1994). The information

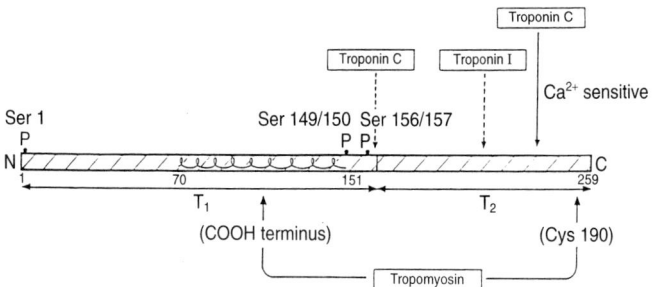

Fig. 4.5. Schematic representation of the structure of troponin T illustrating the sites of interaction with other myofibrillar proteins. Reproduced from Perry (1994).

currently available implies that unlike the situation with troponin I, a single site is involved. This probably involves the remaining strongly conserved region lying between calcium binding sites III and IV, residues 126–132, in the protein from rabbit fast skeletal muscle. This conclusion is supported by the fact that a cleavage product of troponin C obtained by digestion with thrombin, consisting of residues 121–159, forms a complex with troponin T. Whatever the precise nature of the interaction with troponin C, that at the C-terminus of troponin T would appear to be of particular functional importance as it is calcium dependent and when it occurs the binding of tropomyosin in this region is disrupted.

There is evidence that troponin I can bind to troponin T although the interaction sites have not been well characterised. There is no information as yet to the functional significance of this interaction and its well-characterised complex formation with troponin C which modulates the interaction between the latter protein and the inhibitory site on troponin I.

Possible mechanisms of troponin-mediated regulation

The asymmetric nature of troponin T and its interaction with an extended part of the tropomyosin molecule indicates that this component of the troponin complex could lie along the I filament in close proximity to possibly three linearly arranged actin subunits (Fig. 4.1). Nevertheless troponin I, which is about half the mass of an actin monomer with which it must interact at a specific site to inhibit the actomyosin ATPase, is only present at every seventh actin monomer. One of the outstanding problems of muscle regulation is to understand how one molecule of troponin I can effectively control the interaction of seven actin monomers with myosin. With purified actomyosin, the addition of tropomyosin alone amplifies the inhibitory activity of one molecule of troponin I. Clearly, tropomyosin is responsible for this potentiation effect without assistance from the other components of the troponin complex. It is assumed that every actin monomer in the I filament has the capacity of interacting with a myosin head. This would appear to be the case as the sites on actin involved in interaction with myosin are exposed in the reconstruction of the F-actin filament using the data obtained from structural studies on the actin crystals (Holmes *et al.*, 1990). Furthermore, as the myosin heads are flexibly attached to the tail of the molecule there will be no rigid requirements regarding the relative positions of the myosin heads and the actin monomers.

Two possibilities for the mechanism of this effect spring to mind. The first is the so-called steric hypothesis to which reference has been made previously. Such a mechanism would require that conformational changes occurring in the troponin complex on binding calcium are transmitted along the whole length of the tropomyosin molecule, about 40 nm. The fact that the tropomyosin molecule alone is able to amplify the inhibitory activity of small peptides derived from troponin I is not easily explained by the steric hypothesis. For example, the synthetic peptide representing residues 104–115 of troponin I possesses inhibitory activity that is amplified by tropomyosin (Talbot & Hodges, 1981). It is difficult to visualise how a 12 residue peptide about 6% of the size of the troponin I molecule could on binding to an actin monomer displace a tropomyosin molecule of weight 66 000 extending over a distance of 40 nm.

A more plausible hypothesis is that the binding of troponin I to the N-terminal sites of a single actin monomer produces a conformational change in the actin molecule. Evidence for such a change comes from proton NMR studies indicating that the binding of troponin I to actin at the N-terminal sites produces changes at the C-terminal region of actin, to which it is not bound. These conformational changes result in the myosin alkali light chain no longer interacting with this region of actin (Grand *et al.*, 1983). Such changes in actin conformation would be expected to produce changes in the packing arrangement in the neighbouring actin monomers in the same chain and probably in those immediately adjacent in the other chain of the filament. These conformational changes in one actin polymer chain would be strengthened by a similar process mediated by the troponin I associated with an actin monomer at the corresponding point in the other chain of the duplex. With the effects being reinforced at every 38.5 nm it is not unreasonable to visualise conformational changes being transmitted along the length of the filament. As a result the troponin I binding sites on all the actin monomers would be modified so that in resting muscle they are unable to interact with myosin, whether directly blocked by troponin I or not. Activation leading to calcium binding would result in displacement of the troponin I bound to every seventh actin in each polymer thread of the filament. This would reverse the conformational changes at the actin monomer to which the troponin I is bound, changes that would be transmitted as described to the neighbouring actin monomers. In effect the binding sites on all the actin monomers now become available to interact with myosin, resulting in activation of the ATPase. The X-ray diffraction data indicate that there is no detectable change in the dimensions of the I filament helix on stimulation of the muscle. Nevertheless, recent image reconstruction studies of Ishikawa & Wakabayashi (1994) indicate that changes do occur in the actin filament on activation. This implies that the conformational changes in the actin monomers must occur within the constraints of the existing helical structure. It would be expected that such conformational changes in the actin monomers would require adjustment of the tropomyosin-binding position on actin. In this way the movement of tropomyosin that has been shown to occur on contraction, using the X-ray diffraction and image reconstruction studies, could be explained.

Such a theory gives tropomyosin a more passive role in regulation and emphasises its structural role in which it would stabilise and strengthen the I filament structure. It could be considered to provide the template on which the actin filament can respond to the conformational changes imposed upon it. The I filament must be able to maintain the same tension as the A filament which derives its strength from long runs of lateral association of the myosin tails. One might seriously question whether an actin filament built up only of globular subunits would be strong enough to maintain the tension of muscle contracting under maximum load unless it is strengthened by close association with the linearly polymerised tropomyosin molecules aligned in each of the two grooves in the filament. The fact that the movement of tropomyosin is the earliest event to be observed in the X-ray diffraction patterns obtained from intact muscle after the rise in intracellular calcium on stimulation, fits in well with the mechanism proposed above. In the absence of tropomyosin, troponin I appears to be effective as an inhibitor of the actomyosin ATPase in a 1:1 ratio with actin. In the light of the speculation expressed above this would suggest that actin monomers are only able to transmit to adjacent monomers the conformational changes resulting from the binding troponin I and its removal, when the filament is stabilised by tropomyosin molecules acting as a kind of template on either side of the F-actin filament. Such a function rather than one specifically associated with troponin-mediated regulation seems more likely in view of the widespread association of tropomyosin with actin in cytoskeletal systems in many cells including smooth muscle where troponin is absent.

The reports that the differences in the binding constant of actin for myosin in the regulated actomyosin MgATPase system, in the absence and presence of calcium, are not as great as might be expected, do not necessarily argue against a mechanism such as that proposed above. Much has still to be learnt about the nature of the binding between actin and myosin and the effect of the regulatory system upon it. Kinetic studies to determine the effect of calcium on the binding constant binding of myosin for actin in regulated actomyosin MgATPase have

Table 4.2. *Proposed molecular changes occurring in the regulatory proteins of the I filament during the contraction–relaxation cycle in striated muscle*

(1) Intracellular Ca^{2+} concentration rises to 10^{-5} M.

(2) N-terminal calcium specific sites of troponin C, sites I and II occupied by calcium. More slowly the magnesium occupying sites III and IV is replaced by calcium.

(3) The binding of calcium at sites I and II of troponin C results in the N-terminal domain acquiring a more open structure with hydrophobic residues exposed to the solvent.

(4) The interaction between troponin I and C is strengthened, possibly as a result of the conformational changes occurring in troponin C.

(5) Troponin C interacts with the inhibitory peptide region on troponin I and displaces actin from this site.

(6) A conformational change occurs in actin with the removal of troponin I from the binding site. It changes from the 'closed' to the 'open' form. Adjacent actin monomers also change to the 'open' form to accommodate the conformational change but still remain in the double helical structure.

(7) Tropomyosin moves as a consequence of the changes in actin conformation.

(8) All actin monomers now in the 'open' form which are able to interact with the myosin head resulting in the cross-bridge cycle and the associated high rate of MgATPase. Tension develops.

(9) As the Ca^{2+} concentration falls during relaxation calcium is lost from sites I and II on troponin C and the changes that occur on stimulation are reversed.

(10) Troponin C reverts to the form present in resting muscle with the N-terminal calcium-binding sites in a more compact structure and the hydrophobic groups unavailable to the solvent.

(11) The troponin C–I interaction is weakened and the troponin I binds actin at the inhibitory peptide site.

(12) Actin with which troponin I interacts reverts to the 'closed' form as do the other actin monomers by reversal of the process described in (6). Tropomyosin moves back to the resting position as a consequence of actin conformational changes.

(13) Actin no longer able to activate the myosin MgATPase, cross-bridge cycle very much slowed down and tension falls.

usually been carried out using soluble, enzymically active fragments of myosin, subfragment 1 and heavy meromyosin. With the former fragment little change in the binding constant was observed, but with heavy meromyosin it was significant but an order of magnitude lower than what would be expected. There are, however, difficulties in extrapolating these results obtained in a solution of low protein concentration to the situation in the myofibril which is a protein gel of high concentration. The concept of strong and weak binding states of myosin for actin also is a complication in interpreting the results of such binding studies. Modulation of the weaker interaction that is involved in regulating the ATPase activity may therefore not greatly affect the overall binding constant for the interaction.

This mechanism is illustrated in the scheme (Table 4.2) incorporating the observations described above. In it is set out a sequence of events involving the I filament proteins when striated muscle is transformed from the resting to the contracted state by a rise in the sarcoplasmic calcium concentration.

5

Regulation of muscle function by phosphorylation

Evidence now exists for the phosphorylation of the majority of the myofibrillar proteins (Table 5.1). Of the major components of the contractile system only in the cases of actin and troponin C are there no reports of phosphorylated forms existing *in vivo*. Specific kinases are involved in the phosphorylation of myosin, troponin T and tropomyosin but evidence also exists for the phosphorylation of sites on many of the proteins by the less specific cAMP dependent, phosphorylase b and protein C kinases. Usually serine is the preferred site although in some cases threonine can be phosphorylated. The effect of phosphorylation is to introduce negative charges at a site which at some time during the functional life of the protein molecule is exposed to substrate and kinase. As the contractile and regulatory cycles depend on complex interactions between the myofibrillar proteins changes in the net charge at specific sites would be expected to modulate these interactions. This could occur if the phosphorylation site were close to the interaction site either by modifying the electrostatic interactions or as a consequence of the conformational changes resulting from the modification. Phosphorylation at a site distal to the interaction site could also induce conformational changes in the protein that are transmitted through the molecule to the interaction site.

Phosphorylation systems involving the myofibrillar proteins fall into two main groups. The first is that which includes the myosin regulatory light chain (the P light chain) and the cardiac isoform of troponin I. In striated muscle these proteins undergo relatively rapid changes in phosphorylation level involving

Table 5.1. *Proteins of the contracto-regulatory system of verte-brate muscle that have been shown to be phosphorylated*

Protein	Muscle type	Enzymes
Long-term phosphorylation (*not modified by activity, not readily reversible*)		
Troponin I	Skeletal	Phosphorylase b kinase cAMP-dependent kinase
Troponin T	Skeletal	Troponin T kinase
Short-term phosphorylation (*modified by activity, readily reversible*)		
Troponin I	Cardiac	cAMP-dependent kinase
Myosin P light chain	Striated, smooth	Myosin light chain kinase
Myosin light chain kinase	Smooth	cAMP-dependent kinase
Phospholamban	Cardiac	cAMP-dependent kinase Ca/calmodulin kinase

kinases and phosphatases in response to physiological activity. The changes are not sufficiently rapid to be completed in a single cross-bridge cycle but phosphorylation levels increase during activity and fall at a relatively slow rate to the resting level after contraction ceases. They represent short-term modulations of the contractile response by changing the sensitivity of the actomyosin MgATPase to calcium. The other systems involving skeletal troponin I, troponin T and tropomyosin are responsible for the long-term effects. In these cases the protein phosphate is in equilibrium with the phosphate pool of the cell but the process is slow and changes in phosphate content are not observed during short periods of contractile activity as occurs with myosin or cardiac troponin I. For a given muscle type at a particular stage of development the degree of phosphorylation tends to be constant but as in the case with tropomyosin may differ markedly between foetal and adult tissue. It is possible that phosphorylation in these cases occurs during myofibrillogenesis before the proteins are built into the structure. Any changes in the properties of the contracto-regulatory system as a result of phosphorylating these proteins will be characteristic of the muscle type and much more permanent than those resulting from the phosphorylation of myosin and cardiac troponin I.

Tropomyosin

When labelled inorganic phosphate is injected into living muscle or incubated with cells or tissue slices the isoforms of tropomyosin become labelled. The extent of phosphorylation varies between species ranging from 50% in the frog leg muscle, in which tissue the effect was first demonstrated (Ribulow & Barany, 1977), to 10% in rabbit skeletal muscle α-tropomyosin. Although the tropomyosin phosphate equilibrates with the phosphate pool of the cell this is a relatively slow process and the extent of phosphorylation does not change rapidly in response to stimulation or intervention with adrenaline as occurs with the P light chain of myosin or cardiac troponin I, respectively. Higher levels of phosphorylation are found in foetal muscle where for example in the rat heart 60–70% of the α-tropomyosin is in the phosphorylated form (Heeley *et al.*, 1982). During postnatal development the level of tropomyosin phosphorylation in cardiac and skeletal muscles falls progressively to the adult level. In both foetal and adult tissues of the rat α-tropomyosin is significantly more highly phosphorylated in heart than in skeletal muscle.

The site of phosphorylation in rabbit α-tropomyosin is serine 283, the penultimate amino acid residue of the polypeptide chain (Fig. 5.1). Tropomyosin is not phosphorylated by the common phosphokinases but by a specific kinase that has been identified in extracts of chick embryos (deBelle & Mak, 1987). The high levels of phosphorylation associated before and immediately after birth suggest that the protein is more highly phosphorylated when active myofibrillogenesis is in progress. As tropomyosin is a dimeric molecule the phosphorylated form will, at physiological pH values, have up to four additional negative charges very close to the C-terminus. As polymerisation of tropomyosin is presumed to occur by head-to-tail interactions the increase in negative charge would be expected to strengthen the interaction with the N-terminus of the neighbouring molecule which has a net positive charge because of the presence of three lysine residues in the N-terminal decapeptide. The spatial distribution of the amino acid residues in the coiled-coil structure of the tropomyosin molecule is such that the negatively charged phosphate groups at the C-terminus could form electrostatic links with positively charged lysine residues at positions 6 and 12 of the adjacent molecule in the linear polymer (Fig. 5.1).

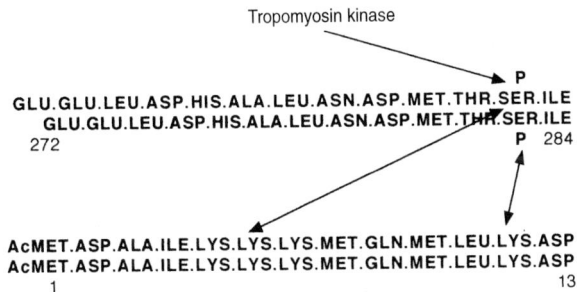

Tropomyosin kinase

P
GLU.GLU.LEU.ASP.HIS.ALA.LEU.ASN.ASP.MET.THR.SER.ILE
GLU.GLU.LEU.ASP.HIS.ALA.LEU.ASN.ASP.MET.THR.SER.ILE
272   P   284

AcMET.ASP.ALA.ILE.LYS.LYS.LYS.MET.GLN.MET.LEU.LYS.ASP
AcMET.ASP.ALA.ILE.LYS.LYS.LYS.MET.GLN.MET.LEU.LYS.ASP
1   13

Fig. 5.1. Sequences of the N and C-terminal regions of the tropomyosin molecule that interact in the linear polymer. In view of the coiled-coil structure of the tropomyosin dimer molecule the negatively charged phosphate residues at the C-terminal end of one molecule interact with the positively charged lysine residues at positions 6 and 12 in the neighbouring molecule. Residue positions indicated by numbers.

Support for such a role is given by the fact that the viscosity of tropomyosin at low ionic strengths, which conditions favour the polymerised form, is higher when the protein is phosphorylated (Heeley *et al.*, 1989; Heeley, 1994). No evidence was obtained in this study for an effect of phosphorylation on the actin-binding properties of tropomyosin but the V_{max} of the actin-activated MgATPase of subfragment 1 in the presence of tropomyosin was increased if the latter protein were phosphorylated. A slightly greater increase was observed if the ATPase was regulated by the troponin system.

It can be concluded that the evidence available indicates that the phosphorylation of tropomyosin is not essential for the regulation of the transduction process but may affect interactions between components of the I filament and thus acts in a modulatory capacity in the response to calcium. The *in-vitro* experiments suggest that by its effect on the actomyosin ATPase, phosphorylation of tropomyosin could have a role in determining the maximum speed of contraction (Heeley, 1994). These studies also suggested that phosphorylation of tropomyosin could modulate its interaction with the N and C regions of troponin T but, surprisingly, in different ways.

Troponin system

Skeletal troponin I

Early studies of the troponin complex isolated from rabbit fast skeletal muscle indicated that preparations contained covalently bound phosphate. When isolated from fresh muscle troponin I contains about 0.5 mole of phosphate and troponin T about 0.7 mole per mole of protein (Perry, 1979). Both these proteins carry a net positive charge at physiological pH values whereas troponin C, which is negatively charged under these conditions, is not phosphorylated.

Two sites have been identified in the fast skeletal muscle

isoform of troponin I from the rabbit that can be phosphorylated *in vitro* and which correspond to the sites that are partially phosphorylated in the intact muscle (Fig. 5.2). These are threonine 11, specific for phosphorylase b kinase, and serine 117 specific for cAMP-dependent protein kinase. These residues are present in homologous positions in the slow skeletal and cardiac isoforms implying that they probably have a function common to all the isoforms. Phosphorylation of the two sites is not an artefact of isolation arising from the action of the endogenous enzymes during preparation. When troponin I is isolated from fresh muscle by an affinity chromatographic method under denaturing conditions which prevent enzyme action (Syska *et al.*, 1974) it is partially phosphorylated. This is despite the fact that in the intact muscle troponin I is complexed with troponin C which inhibits phosphorylation at both sites *in*

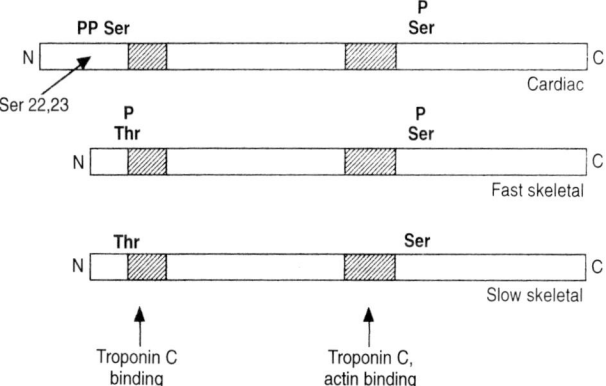

Fig. 5.2. Schematic representation of the isoforms of troponin I from vertebrate striated muscle. Troponin C and actin bind in the region of the inhibitory peptide which consists of residues 96–116 in the rabbit fast skeletal muscle isoform. In this protein the phosphorylation sites are threonine 11 and serine 117. Shaded areas represent regions of homology in the isoforms.

vitro. The inhibition is not surprising for both phosphorylation sites are in regions of the molecule at or very close to those known to interact with troponin C. The nature of the troponin I–C interaction clearly must change during the contractile cycle (Chapter 4) but no significant change in the phosphorylation state of skeletal muscle troponin I can be demonstrated as a consequence of continued contractile activity in tetanus (Ribulow *et al.*, 1977). It seems likely that phosphorylation occurs at some stage during biosynthesis before troponin I is incorporated into the troponin complex. Covalent modification close to the sites of interaction would be expected to modulate the properties of the complex in the myofibril. Both of the regions of troponin I involved in the interaction with troponin C are positively charged because of the presence of arginine residues, suggesting that there is a significant electrostatic component to the interaction. Phosphorylated troponin I will still form a complex with troponin C but the balance of charges at the interaction sites would be changed by the introduction of negative charges close to the regions involved. This would be expected to modify the interaction between the two proteins and presumably affect the calcium-binding properties of troponin C.

Cardiac troponin I

When isolated from fresh tissue the cardiac troponin complex contains significantly larger amounts of covalently bound phosphate than does the complex obtained from skeletal muscle. This is accounted for by the much higher phosphate content of cardiac troponin I because of the presence of an additional phosphorylation site in this protein (Fig. 5.2). The amino acid sequence of cardiac troponin I exhibits about 60% homology to the slow and fast skeletal muscle isoforms with the regions involved in the interaction with troponin C and actin virtually identical in all three isoforms. The major sequence difference is an additional N-terminal peptide of 29–30 residues in which positions 23 and 24 in the cow and human are occupied by serines that are readily phosphorylated by protein kinase (Mittmann *et al.*, 1990). Originally, it was assumed from the published sequence that there was only one serine at the phosphorylation site in the N-terminus. The difficulties of determining the amounts of phosphoserine present were not appreciated in the original primary structural studies (Moir & Perry, 1977). *In vitro* the N-terminal site and serine 146 (homologous to serine 117 of the fast skeletal isoform in the rabbit) are both phosphorylated by protein kinase but the N-terminal site much more rapidly. When the rabbit heart is perfused with ^{32}P labelled inorganic phosphate the latter site is the only one that becomes significantly labelled indicating that it is in rapid dynamic equilibrium with the intracellular pool of ATP. On treatment with adrenaline the phosphorylation of cardiac troponin I increases up to 2 moles per mole. The increase observed in the original studies was confined to the N-terminal region, residues 1–44, and it was originally considered, for the reasons explained above, that a site additional to the originally identified protein kinase site (wrongly assigned from the sequence studies of Grand *et al.* (1976) to serine 20) had become labelled. This additional site could not be identified and the suggestion that one existed is no longer valid as it has been shown (Mittmann *et al.*, 1990) that both serines in positions 22 and 23 of rabbit cardiac troponin I can be phosphorylated and thus account for the increase in phosphate in the N-terminal region.

In vitro serine 22 and 23 of rabbit cardiac troponin I are phosphorylated by cAMP-dependent protein kinase and it is considered that this enzyme is activated after intervention with adrenaline *in vivo* to convert the troponin I to the bisphosphorylated form. It is not possible to say precisely at this stage how the phosphate is distributed between the two serine residues in the heart before intervention with adrenaline. The values obtained for total phosphate in these conditions are usually slightly in excess of one mole per mole which, if allowance is made for possible limited phosphorylation of minor sites such as serine 149, could be interpreted to imply that one of the serines at the N-terminal site is fully phosphorylated. The limited evidence currently available, however, suggests that both monophosphorylated forms may be present in untreated fresh hearts (Mittmann *et al.*, 1990). Studies with synthetic peptides corresponding to the N-terminal region of cardiac troponin I indicate that with cAMP-dependent protein kinase ordered phosphorylation occurs at the double serine site (Mittmann *et al.*, 1992; Quirk *et al.*, 1995). In the case of the human isoform, phosphorylation of serine 24 is virtually complete before serine 23 is phosphorylated. Proton NMR studies show that diphosphorylation induces a marked conformational transition because of a much increased interaction between the phosphoserine 24 and arginine 22 (Quirk *et al.*, 1995). On the basis of these studies it has been suggested that the transition from the mono to the diphosphorylated form of cardiac troponin I is the important step in decreasing the calcium sensitivity of the system. Recent studies with phosphorylated mutants of cardiac troponin I containing single serines at the phosphorylation site confirm that both serines must be phosphorylated to produce the change in calcium sensitivity (Zhang *et al.*, 1995).

It is now clearly established that the phosphorylation of troponin I resulting from the intervention with adrenaline produces a fall in the sensitivity of the actomyosin MgATPase to calcium, that is, the calcium concentration required for 50% activation increases (Ray & England, 1976; Solaro *et al.*, 1976). A consequence of this effect is an increase in the speed of relaxation that is a feature of the action of adrenaline on the heart. Two adjacent serine residues are present in homologous positions in the N-terminal peptide of rabbit, rat, quail, ox and human cardiac troponin I suggesting that modulation of calcium sensitivity by phosphorylation of this protein is a general feature of mammalian cardiac metabolism. Thus by modulating the response to calcium through the influence of an external factor on troponin I (and on phospholamban, see later) the contractile force developed by every cardiac cell can be varied. This is an important property of the myocardium but not of skeletal muscle which varies the force developed by recruitment of fibres.

The mechanism of this effect and the relative roles of phosphorylation of each of the two N-terminal serine residues is of some interest. In cardiac muscle contraction is controlled by calcium binding at site 2 of troponin C as site 1 does not function as a high affinity site (Chapter 3). The effect of increasing the net negative charge at the two N-terminal serines of the troponin I could be transmitted to the calcium-binding domain of troponin C to explain the enzymic effects. In the case of skeletal troponin I the phosphorylation sites are located at the regions known to be involved in interaction with troponin C. Nevertheless, so far there has been no evidence of changes in the phosphorylation levels of threonine 11 and serine 117 occurring during activity in skeletal muscle and the role of phosphorylation at these sites is a matter for conjecture. Clearly they do not have a dynamic role. The corresponding sites in cardiac troponin I do not appear to have any role in the response to adrenaline intervention.

The N-terminal serine site on cardiac troponin I appears from NMR studies to be on the surface of the molecule and to be relatively mobile. Troponin C does not block its phosphorylation either *in vitro* or *in vivo* where the extent of phosphorylation varies as a physiological response. This would suggest that the phosphorylation of the N-terminal serines exerts its effect by inducing a conformational change in troponin I that is conveyed through the interaction sites with troponin C. Another possibility is that the flexible N-terminal site is able to interact directly with troponin C in some manner not yet understood but which leaves the site available for phosphorylation. It would appear more likely that the phosphorylation of troponin I exerts it effect directly on troponin C, but other possibilities cannot be excluded. For example, the conformational change in the troponin I induced by phosphorylation might act directly on its interaction with actin in a manner that indirectly reduces the calcium sensitivity of the system.

Troponin I contains a number of sites other than those to which reference has been made above that are potentially available to phosphorylation by the kinases present in the muscle cell. When bovine cardiac troponin I for example is incubated with protein kinase C between one and two moles of phosphate are incorporated, principally at threonine 199 and threonine 280 (Noland & Kuo, 1993). This results in a fall in the V_{max} of the calcium-activated MgATPase but no change in calcium sensitivity. It is suggested that the fall in activity results from an effect on the interaction between the thick and thin filament proteins. These results were obtained from *in-vitro* phosphorylation studies but the authors consider they may explain the negative inotropic effect of phorbol ester on cardiac muscle. As yet there is no evidence of the *in-vivo* phosphorylation of the sites responsible for this effect.

Troponin T

Troponin T contains the major part of the covalent phosphate present in the troponin complex isolated from rabbit skeletal muscle. When isolated the troponin T contains about 0.7 mole of phosphate per mole, most of which is located at the N-terminal serine. *In vitro* troponin T is not a significant substrate for protein kinase but can be phosphorylated by phosphorylase b kinase. With the latter enzyme, up to 3 moles of phosphate per mole can be introduced but at an initial rate which is only about 5% of that obtained with its normal substrate. Under these conditions three phosphorylation sites can be identified, serine

1, serine 149–150 and serine 156–157 (Fig. 4.5). Serine 1 is the major site of phosphorylation when troponin T is isolated and for which a specific enzyme, troponin T kinase, has been isolated (Gusev *et al.*, 1980). This would suggest that it has functional significance but as yet there is no information as to what it might be.

As is the case with skeletal troponin I, contractile activity does not change phosphorylation levels of troponin T. Likewise troponin C in approximately equimolar proportions markedly inhibits phosphorylation of all three sites (Moir *et al.*, 1977). This would suggest that the phosphorylation sites are blocked in the complex with troponin C. It is somewhat surprising that phosphorylation at sites widely separated in the primary sequence should be inhibited in view of the evidence that troponin T exists as an extended molecule.

Despite a slightly longer polypeptide chain cardiac troponin T is highly homologous with the skeletal muscle isoforms. When isolated it usually contains less covalent phosphate than the skeletal troponin T. It is uncertain if this is an artefact of preparation for the phosphorylation sites are homologous, namely serine 1 and serine 176 (equivalent to the serine 150 site in the skeletal isoform; Fig. 4.5). Nevertheless there are differences in the properties of the two proteins in that the phosphorylation rate with phosphorylase b kinase is slower with the cardiac form and it is not inhibited at either site by troponin C (Raggi *et al.*, 1989).

Myosin

The phosphorylation of myosin is a dynamic process that occurs to varying degrees as a consequence of stimulation in all types of vertebrate muscle. The enzyme responsible, myosin light chain kinase, phosphorylates a serine residue close to the N-terminus of the phosphorylatable light chain (P light chain), sometimes called the regulatory light chain (Fig. 5.3). The latter terminology is often used to distinguish its function from that of the other light chain which is known as the essential or alkali light chain 1 (Chapter 2). As neither light chain is essential for the ATPase activity of vertebrate myosin and the evidence that is accumulating suggesting that both may play a regulatory role in the function of the myosin motor, it will be referred to here as the P light chain in those muscles with an active kinase.

Myosin light chain kinase, which was first identified and purified from rabbit fast skeletal muscle (Pires & Perry, 1977), has been shown to be widely distributed in muscle and non-muscle tissues. A striking exception is molluscan adductor muscle in which the enzyme is not present at a detectable level. Nevertheless, adductor myosin contains a light chain homologous to the P light chain of vertebrate striated and smooth muscle myosins. The regulatory light chain of molluscan adductor is not phosphorylated *in vivo* but it can act as a

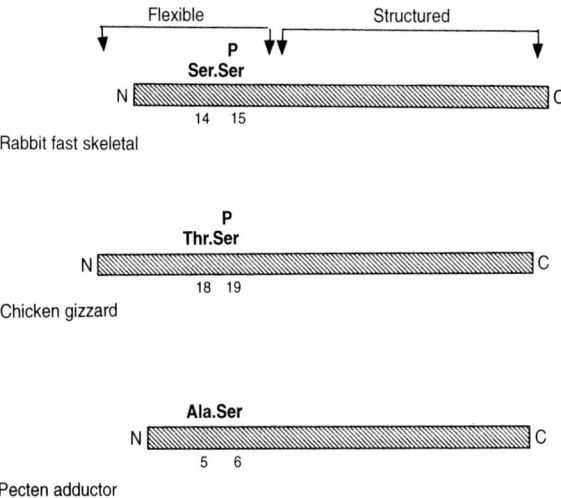

Fig. 5.3. Schematic representation of the regulatory (phosphorylatable) light chains of myosin. Residue positions indicated by numbers.

substrate for the light chain kinases from vertebrate tissues *in vitro*. The myosin light chain kinases from different species and tissues are active as monomers but their molecular weights vary considerably. In general the smooth muscle enzymes are larger, for example that from avian gizzard which has been widely studied has a molecular mass of about 130 kDa. A similar value has been reported for chicken skeletal muscle whereas lower values, in the range of 70 to 90 kDa, are obtained for mammalian skeletal and cardiac forms. These are apparent molecular weights obtained by electrophoresis which appear to give anomalous results with these proteins. The molecular mass of rabbit fast skeletal enzyme, originally reported to be 77 kDa, was shown from the amino acid sequence to be 65 kDa.

Calmodulin and calcium are essential for the activity of the enzyme. The mechanism is similar to that proposed for other calmodulin-dependent enzyme systems in which probably all four of the calcium-binding sites on calmodulin are occupied. In this state it forms a one-to-one complex with the kinase and activates the enzyme.

$$4 \, Ca^{2+} + calmodulin \rightarrow Ca^{2+}{}_4. \, calmodulin$$
$$Ca^{2+}{}_4.calmodulin + kinase \rightarrow Ca^{2+}{}_4. \, calmodulin.kinase$$

The kinase molecule is characterised by a highly asymmetric N-terminal region with a low α-helix and a high proline content (Fig. 5.4). In the mammalian skeletal muscle enzyme this region, representing almost one-half of the molecule, is much more variable than the C-terminal moiety. For example, the sequences of the N-terminal regions of the rabbit and rat skeletal muscle enzymes are about 58% homologous. This domain, which is responsible for the anomalous electrophoretic behaviour, is of unknown function. The C-terminal region in the rabbit skeletal enzyme contains the calmodulin-binding site, consisting of about 30 residues at the C-terminus, and the catalytic domain. Within the catalytic domain is a peptide region that resembles the sequence about the phosphorylation site in the P light chain, particularly with respect to the distribution of the basic residues in relation to the serine residue that is phosphorylated. It has been suggested that in the inactive form of the enzyme this region interacts with the active site and renders it ineffective. In the presence of calcium and calmodulin this region is displaced and the site becomes catalytically active (Fig. 5.4). The importance of the C-terminal domains for the function of the enzyme is indicated by the fact that in this region of the rabbit and rat skeletal muscle enzymes there are only 11 amino acid differences in sequence, all but two of which are conservative.

Myosin light chain kinase has a narrow substrate specificity, the preferred phosphorylation site being a serine residue close to the N-terminus of the P light chain, serine 15 in rabbit fast skeletal muscle myosin (Fig. 5.3). The corresponding site in the P light chain of chicken gizzard myosin is serine 19. In the latter case a threonine residue in position 18 can also be phosphorylated but at a very much slower rate than serine 19 and

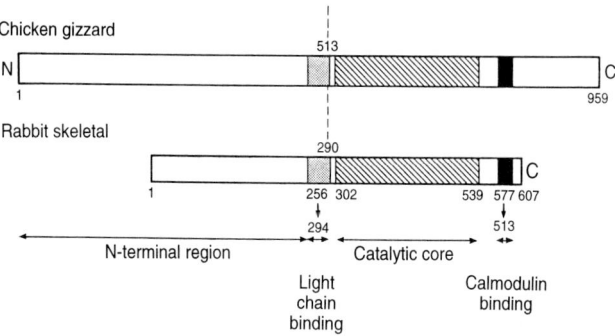

Fig. 5.4. Schematic representation of striated and smooth muscle myosin light chain kinases. Vertical dotted line indicates residues at which sequence alignment can be made to give 11 invariant residues in the catalytic core (Knighton *et al.* (1992). The least homology is exhibited in the N-terminal regions which are polar and rich in proline.

requiring a large excess of kinase to achieve significant phosphorylation *in vitro*.

Phosphorylation of threonine 18 does not appear to be required for the activation of myosin ATPase and its biological significance, if any, is not clear. It should be pointed out that the N-terminal region of the P light chain in which the phosphorylation site is located is a very flexible part of the molecule, so much so that the location of the amino acid residues in it could not be precisely defined in the X-ray structural studies of subfragment 1 (Rayment *et al.*, 1993).

The specificity of the enzyme for myosin and the fact that phosphorylation rises as a result of stimulation implies its involvement in muscle function. In the case of striated muscle the enzymic activity is highest in fast skeletal, lower in slow skeletal and very low in cardiac muscle. Immunochemical evidence suggests that the kinase exists as muscle type-specific isoforms implying that the enzyme has evolved in response to the development of the specialised physiological properties of the different muscle cell types as is the case with the myofibrillar proteins. This in itself suggests that the enzyme has a functional role.

As the enzyme is readily extracted from striated muscle by low ionic strength buffers it is presumably dissolved in the sarcoplasm and evenly distributed in the cell. On this assumption the maximum rate of myosin phosphorylation that can be expected, even in fast skeletal muscle which contains the highest amount of kinase, is only 1–2% of that of the actomyosin MgATPase. Clearly myosin phosphorylation is not a required step for the cross-bridge cycle but in some way modulates the process. Kinetic studies on skeletal actomyosin MgATPase systems indicate that phosphorylation of myosin results in a fall in the K_m, implying an increase in affinity for actin. The V_{max} is not affected which correlates well with the reports that phosphorylation does not affect the maximum shortening velocity or force production in skinned fibres. In resting skeletal muscle the P light chain is 10–50% phosphorylated, the value depending on the muscle and species. The rate of increase of myosin phosphorylation depends on the frequency of stimulation and not the total number of stimuli applied. For example, stimulation *in situ* of combined rabbit fast muscles, extensor digitorum longus and tibialis anterior, for prolonged periods at 1 Hz failed to produce a marked increase in phosphorylation above resting levels (Westwood *et al.*, 1984). In contrast, stimulation at 4 Hz for 30 s resulted in 80% phosphorylation of the P light chains. Presumably the effective steady-state calcium level is higher at the higher frequencies of stimulation and results in higher kinase activity.

Potentiation of isometric twitch tension after tetanus is a well-known property of fast skeletal muscle, the so-called post-tetanic potentiation. There is a correlation between this effect and the increase in P light chain phosphorylation (Manning & Stull, 1979) at the higher levels of phosphorylation. The correlation is not, however, complete in rabbit fast muscle for phosphorylation persists at 40–50% above the resting value for some time after the potentiation effect has disappeared (Westwood *et al.*, 1984). The relatively slow rate of myosin dephosphorylation is a reflection of the low levels of myosin light chain phosphatase activity in striated muscle. For example, 16 min after a 5 s tetanus the phosphorylation of myosin in the combined extensor digitorum longus and tibialis anterior muscles of the rabbit was still 20–30% above the resting value.

Another possible role of phosphorylation has been suggested by the observation with skinned skeletal muscle fibres that myosin phosphorylation increases the sensitivity of the contractile response to calcium at suboptimal concentrations of the cation (Sweeney & Stull, 1990). This implies that some mechanism of calcium regulation directly involving myosin may exist in vertebrate striated muscle, evidence for which was also provided some years ago by Lehmann (1978). Nevertheless, the bulk of evidence available suggests that changes in calcium sensitivity in vertebrate skeletal muscle are mediated principally through the troponin complex. Using the analytical framework of Brenner (1988) to interpret the response of permeabilised rabbit psoas fibres to myosin phosphorylation, Sweeney & Stull (1990) concluded that increasing myosin phosphorylation from less than 10% to more than 75% increases the rate constant describing the transition of the non-force-generating cross-bridges to the force-generating state. As the rate constant of the reverse process is not changed by phosphorylation this conclusion could explain the increase of twitch tension observed when myosin is phosphorylated.

Recently Tohtong *et al.* (1995) have demonstrated the importance of myosin light chain phosphorylation for the stretch activation response of the indirect flight muscles of *Drosophila melanogaster*. Insect indirect flight muscles are similar in structure to vertebrate striated muscle but in addition to calcium they also require to be stretched for full activation. Mutants were prepared in which the light chain phosphorylation site of the P light chain, serines 66 and 67 (corresponding to serines 15 and 16 in the rabbit), were substituted. Myofibrillogenesis and the development of maximal isometric tension was unimpaired but the mutant insect had reduced flight ability and the isolated fibres had a reduced power output.

The myosin light chain kinase activity of slow skeletal muscle is much lower than that of fast muscle. In consequence the extent of myosin phosphorylation is less in the resting muscle and prolonged periods of tetanic stimulation at frequencies above the natural rate of stimulation are required to raise significantly the myosin phosphorylation level. The P light chain fraction of myosin from rabbit soleus muscle consists of two isoforms, both of which are phosphorylated. In contrast to the fast form, slow skeletal muscle does not exhibit post-tetanic potentiation (Close & Hoh, 1969) which suggests that myosin phosphorylation may not be the only factor associated with this effect. Heart myosin resembles that of slow skeletal muscle in containing two isoforms of the P light chain, both of which exist in the phosphorylated form (Westwood & Perry, 1982). The kinase level in cardiac muscle is very low and although in the normal functioning heart of the rabbit, for example, the myosin is about 25% phosphorylated there is a barely significant rise when activity increases, for example after exposure to adrenaline. Cardiac ischaemia also produces a slight increase in phosphorylation.

So far as striated muscle is concerned it must be concluded that there is still some uncertainty as to the physiological role of myosin phosphorylation. The kinetic evidence indicating an increase in affinity of myosin for actin when phosphorylated is compatible with the correlation between the potentiation of twitch tension after tetanus and the rise in phosphorylation of the P light chain. The increase in calcium sensitivity at low calcium concentrations is of particular interest in view of the role of the regulatory light chain of molluscan myosin in conferring calcium sensitivity to molluscan actomyosin ATPase and may imply a more general role for the P light chain in modulating the calcium sensitivity of the actomyosin systems.

The fact that myosin possesses two identical enzymically active heads each with a phosphorylatable light chain raises the question of the functional relationship of the two heads when they are covalently modified. Early claims of ordered phosphorylation of the two heads have not been confirmed, the current evidence suggesting that the process is random. It is not clear, however, whether both heads of myosin have to be phosphorylated for the physiological response to occur or whether each head can respond independently to the phosphorylated state.

General comments on myosin phosphorylation

Despite the homology between the sites of phosphorylation in all vertebrate myosins it is remarkable that the physiological consequences of the covalent modification do not appear to be the same in different muscle types. Two effects of phosphorylation of the P light chain on the actin-activated MgATPase, on which the cross-bridge cycle and tension development depend, can be distinguished. One effect is the increase in rate which is very marked with smooth myosin but much less so with skeletal muscle myosin where it is only observed under certain conditions. The other is the increase in calcium sensitivity reported for the skeletal system in conditions of suboptimal calcium concentration. It is not clear how these effects are related because the calcium sensitivity of the MgATPase in striated muscle is usually considered to be determined by the binding of calcium to troponin C and there is no information as to how the phosphorylation of the P light chain could influence this system. It may be that the myosin light chains have a role in determining the calcium sensitivity of the MgATPase in vertebrate striated muscle by direct effects on the myosin molecule but this is only apparent under conditions of limiting calcium concentration. A role in calcium-mediated regulation is well documented in the more primitive molluscan adductor muscle. It is possible that during evolution and specialisation this property still survives in an attenuated form in the much faster actin-activated ATPase of vertebrate skeletal muscle. An association of calcium regulation with the myosin motor itself is suggested from study of the unconventional myosins, many of which have calmodulin-binding sites.

Calcium sensitivity of molluscan actin-activated MgATPase depends on the presence of the regulatory light chain, which is homologous with the P light chain of vertebrate muscle. In its absence the actin-activated ATPase is high regardless of the calcium concentration. When the regulatory light chain is restored to the myosin the ATPase is low in the absence of calcium but becomes activated in its presence, that is, the ATPase becomes calcium sensitive (Kendrick-Jones *et al.*, 1970). It has recently been shown (Xie *et al.*, 1993) that a novel EF hand calcium-binding domain on the so-called essential light chain is stabilised by linkages with the heavy and both light chains (Plate 2.1(*b*)). Thus when calcium is bound to this novel site the inhibitory effect of the regulatory light chain is removed. One can presume that the binding of calcium to the site induces conformational change in some part of the myosin molecule that enables it to interact with actin in the normal manner that leads to activation of the MgATPase. In all types of actomyosin contractile systems it is very likely that regulation involves controlling this aspect of the interaction. This interaction is relatively weak and sensitive to ionic strength, at least when observed in *in-vitro* systems (Chapter 2). In the more primitive contractile systems, such as molluscan adductor and vertebrate smooth muscle, regulation of this interaction involves changes in the myosin molecule itself. The regulatory and P light chains inhibit the actin-activated ATPase in molluscan adductor and vertebrate smooth muscles, respectively, by maintaining the myosin in a conformation that does not permit actin to activate the ATPase. In the most primitive smooth muscle, the molluscan adductor, binding calcium at the novel site relieves this inhibition as explained above. In the more advanced vertebrate smooth muscle the P light chain maintains its inhibitory role which is relieved when it is phosphorylated. The activating effect of calcium is now on the kinase introducing an additional step in the process and thus increasing the possibility of modulating the contractile response, an important aspect of vertebrate smooth muscle function. It follows that binding calcium at the novel binding site directly involving the two light chains, neither of which is phosphorylated, and which are located in the mid-region of the helical neck of the molluscan myosin head should have the same effect as phosphorylating a site on the P light chain located closer to the distal end of the neck region in vertebrate smooth myosin (Plate 2.1(*b*)). Both events involve changing the charge balance at the sites involved and it is possible that similar conformational changes are induced which are adequate to displace the light chain from its inhibitory conformation.

In addition to its role in regulating the interaction with actin and hence the ATPase, the phosphorylation state of the light chain in smooth and non-muscle cells determines the assembly of myosin into filaments. Under physiological conditions non-phosphorylated smooth muscle myosin exists as a monomer in folded conformation with a sedimentation coefficient of 11S. On phosphorylation the molecule unfolds to the extended monomeric form with the conventional sedimentation coefficient of 6S. This form is able to assemble into filaments (Smith *et al.*, 1983).

One may speculate as to whether the P light chain in vertebrate muscle has retained any of the calcium-sensitising properties that are associated with its homologue in the primitive molluscan system. In view of the points made above this would appear to be probable in smooth muscle although direct evidence for this is not yet available. Such a property could explain the occasional reports in the literature of high ATPase activity of actomyosin systems and tension development under conditions when the calcium level is high but which do not correlate with a high level of P light chain phosphorylation.

In contrast to smooth muscle in which calcium-mediated regulation acts principally via the myosin filament, the target for calcium in striated muscle is the troponin complex located in the thin filament. This system has evolved to enable striated muscle to respond rapidly to a stimulus. In effect binding calcium on a single molecule of troponin C results in an amplification effect that results in seven actin monomers being made available to interact with myosin. Rapid regulation via the thin filament in this manner requires that all the myosin heads are in the correct conformation to interact with actin in a manner that results in high MgATPase activity. To accommodate this requirement the P light chain no longer has a marked inhibitory activity as is the case in vertebrate smooth muscle and with its homologue, the regulatory light chain, in molluscan adductor muscle. In striated muscle the inhibitory component, troponin I, acts directly on actin to prevent its activation of the myosin ATPase. Nevertheless, in view of the effects of light chain phosphorylation reported for striated muscle (Sweeney & Stull, 1990) it is not unreasonable to assume that the covalent modification results in conformational changes in the myosin head that lead to changes in the actomyosin MgATPase and the direct effects of calcium on it.

Phospholamban

The reversible phosphorylation systems of cardiac troponin I and myosin modulate the response of the myofibrillar actomyosin ATPase to calcium during contractile activity. A phosphorylation system with many similarities in which the protein substrate is phospholamban regulates the calcium pump of the SR and effects the contractile response by modulating the calcium transients of the muscle cytoplasm.

Phospholamban is a 52 amino acid residue protein associated in pentameric form with the calcium pump ATPase in the membrane of the SR. When the myocardium is treated with adrenergic agents phospholamban is phosphorylated with an associated increased rate of calcium transport across the SR. This leads to an increased accumulation of calcium in the SR which in turn leads to an increased rate of release. Thus the enhanced calcium-induced calcium release and the increased rate of calcium loading of the SR could explain the increased contractility and contribute to the speeding up of relaxation (abbreviation of systole), respectively, resulting from the action of adrenaline.

The phosphorylation site on phospholamban consists of serine 16, a substrate for the cAMP-dependent kinase, and threonine 17 which is phosphorylated by a calmodulin-dependent protein kinase. As is the case with the N-terminal site of cardiac troponin I, the relative functional significance of the two sites is not clear. Likewise, the N-terminus is a motile region of the phospholamban molecule (Fig. 5.5). The phosphorylation site lies at the C-terminal end of a short run of α-helix, which is joined up to the bulk of the molecule that is hydrophobic and embedded in the membrane, by a non-helical flexible region. It is considered that phospholamban exerts its effects on two rate-limiting steps of the calcium ATPase cycle by acting as an inhibitor in its unphosphorylated form (for a review, see Tada & Kodama, 1989). Phosphorylation relieves the inhibition and thus speeds up calcium transport into the SR. The steps involved are stages 3 and 5 in the generally accepted scheme for the mode of action of the calcium pump (Chapter 3). These steps are those in which the major conformational changes in the ATPase are considered to occur. A mechanism can be visualised in which

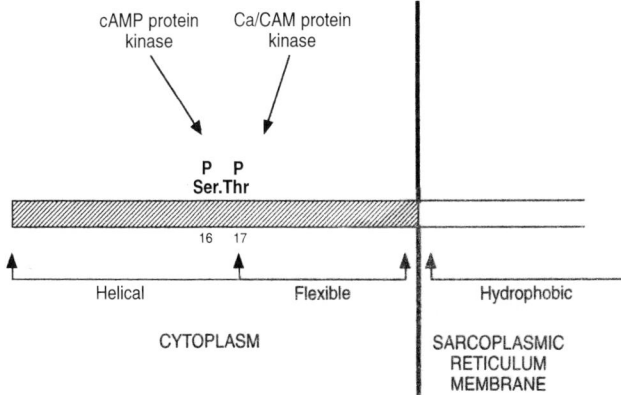

Fig. 5.5. Diagrammatic representation of the N-terminal region of phospholamban. The protein is associated with the calcium pump ATPase in the membrane of the sarcoplasmic reticulum. The flexible N-terminus on which the phosphorylation sites are located extends out into the cytoplasm of the cardiac muscle cell. Numbers indicate residue positions. CAM, calmodulin.

interaction between the two proteins is modified in such a way that when phospholamban is phosphorylated the affinity for calcium at the binding domain on the ATPase is increased.

Common features of reversible phosphorylation systems in functioning muscle

Comparison of phospholamban phosphorylation with the two other physiologically reversible phosphorylation systems present in muscle that are known to modulate calcium response indicates some remarkable common features in all three systems:

(1) The protein that is the phosphorylation target is of relatively low molecular weight.
(2) The protein that is phosphorylated possesses inhibitory activity on a transducing ATPase.
(3) The inhibitory protein has a flexible N-terminal region in which is located the phosphorylation site consisting of two adjacent HO amino acids.
(4) Phosphorylation results in changes in the calcium-binding properties of an interacting protein.

In the light of these common properties it is conceivable that a basically similar mechanism of action applies in all cases. It would appear that this type of molecular arrangement involving reversible phosphorylation is very suitable for modulating the sensitivity to calcium of the contracto-regulatory system of muscle. These systems differ from the phosphorylation systems presumed to be responsible for long-term modulation to calcium. In the latter the sites are usually single and located at or close to the peptide regions known to be involved in interactions between the protein components of the regulatory system. Consequently, in the intact functioning muscle most of these sites would not be available to the specific kinases involved in their phosphorylation.

6

Dystrophin and the muscular dystrophies

The term muscular dystrophy, first introduced by Erb (1891), is applied by convention to a number of different muscle disorders, the most important of which are listed in Table 6.1. These disorders have certain features in common, namely a hereditary nature, primary involvement of the voluntary muscle and a tendency to progressive deterioration as a result of gradual and piecemeal necrosis. Their frequency of occurrence is low with Duchenne muscular dystrophy (DMD) accounting for the largest group of patients. Within the dystrophies there is a wide range in the severity of the condition and the expectation of life of the patients. Often the pattern of muscle involvement is highly selective, particularly in the early stages of the disease. This is a puzzling aspect of these conditions as one would expect a genetic defect to be equally severe in all muscle cells. A possible explanation is that the activity pattern, which will be characteristic of the muscle, in some cases determines the susceptibility of the muscle to the genetic deficiency.

In addition to the dystrophies listed in Table 6.1 there are other groups of inherited conditions that lead to varying degrees of muscle malfunction. One group includes the myopathies in which, from the nature of the biochemical changes associated with the condition, it is often clear as to the gene that has undergone mutation. These include conditions in which there are deficiencies in the enzymes involved in carbohydrate and lipid metabolism and in the oxidative pathways, as is the case in the rapidly expanding group of mitochondrial myopathies (Table 6.2).

Table 6.1. *Major muscular dystrophies*

Dystrophy	Frequency (per 10^5)	Chromosome location
Duchenne	3 (recessive)	Xp21
Becker	1 (recessive)	Xp21
Emery—Dreifuss	1 (recessive)	Xq28
Myotonic	5 (dominant)	19q13
Fascioscapulohumeral	< 1(dominant)	4q35
Ocularpharyngeal	(dominant)	Autosome
Limb girdle syndromes		
SCARMD	(recessive)	13q
Limb girdle	(recessive)	15q
Late-onset limb girdle	(dominant)	5q

SCARMD, Severe childhood autosomal recessive muscular dystrophy.

The conditions described above are true myopathies in which the genetic lesion is expressed in the muscle cell but degeneration of muscle can also occur as a consequence of diseases of the motor neurone. For example, the spinal muscular atrophies represent a complex group of inherited disorders in which the muscle atrophies because of degeneration of the lower motor neurones. The resulting muscle weakness leads to deformation of the skeletal system as is the case with DMD.

In the cases of the metabolic disorders listed in Table 6.2 the biochemical changes observed in the diseased muscle have enabled the gene involved to be identified. This approach has had limited success so far as the dystrophies are concerned probably because the genes concerned, in some cases at least, are responsible for the synthesis of proteins involved in aspects of cell chemistry as yet not defined. Changes in the cell chemistry have been observed, particularly in DMD. These are often secondary to the cause of the disease and may arise because of the changes

Table 6.2. *Hereditary myopathies in addition to the muscular dystrophies*

Condition	Mutant protein	Chromosome
Glycogenoses		
Pompe, Type II glycogenosis	Acid maltase	17q23–25
Cori–Forbes, Type III glycogenosis	Debrancher enzyme	1p21
Andersen, Type IV glycogenosis	Branching enzyme	—
McArdle, Type V glycogenosis	Phosphorylase	11q13
Tauri, Type VII glycogenosis	Phosphofructokinase	1
Type VIII glycogenosis	Phosphorylase b kinase	X(α), 16(β), 7(χ)
Type X glycogenosis	Phosphoglycerate kinase	Xq13
Type X glycogenosis	Phosphoglycerate mutase	7,10
Type XI glycogenosis	Lactic dehydrogenase	11,12
Disorders of lipid metabolism		
Carnitine deficiency	Carnitine palmitoyltransferase	
	Defective carnitine transport	
Mitochondrial diseases	Factors involved in oxidative metabolism	Mitochondrial genome
Malignant hyperthermia	Ryanodine receptor	

of the muscle phenotype that can occur as a consequence of the changed activity pattern resulting from the dystrophy. Also they may reflect the regenerative response that is characteristic of some dystrophic conditions or the general degenerative changes that occur as a consequence of the genetic lesion. These changes often throw little light on the lesion itself that has triggered off the modifications to the cell chemistry.

An alternative approach adopted to discover the lesion responsible has been to attempt to identify the gene without prior knowledge of its function, originally called 'reverse genetics' or more recently 'positional cloning'. The procedure involves determination of the chromosome involved followed by identification of the region of the chromosomal DNA in which the lesion occurs. This is obtained by linkage studies with probes and in some cases from structural changes on the chromosome associated with the condition. The gene is cloned and by isolation of the appropriate messenger RNA the sequence of the cDNA is determined and the amino acid sequence of the gene product obtained. Such an approach has lead to the identification of the gene involved in DMD and its success has stimulated similar research on the other dystrophies. The result has been that the chromosomes involved in most of the conditions listed in Table 6.1 have recently been identified and progress is now well under way to locate the genes concerned.

Duchenne muscular dystrophy

The physician, Duchenne, after whom the disease was named published his observations in 1861. The condition, however, had been reported earlier by Little (1853) who described two brothers with the characteristic symptoms of the condition. DMD has an incidence of approximately 1 in 3500 male births with about one-third of the cases considered to arise by spontaneous mutation. As one of the most common forms of muscular dystrophy it occurs in a relatively homogenous group of patients with consistent symptoms. It is virtually confined to boys as the condition is transmitted by X-linked recessive inheritance. The severe form of the disease occurs in females extremely rarely as a consequence of X:autosomal translocation or a failure of inactivation of the mutant form of the maternal X chromosome.

Before the gene probes became available DMD was not easily diagnosed in early infancy where it is characterised by developmental delays and difficulties in climbing and running. Between 3 and 6 years of age a characteristic gait is apparent and from 6 to 11 years the strength of the muscles decreases steadily. Ambulation is lost by the second decade and death occurs by or soon after the end of the second decade. Since 1953 a more benign form of the disease, Becker muscular dystrophy (BMD), has been recognised. This form exhibits many of the symptoms associated with DMD, for example pseudohypertrophy of the calf muscles and proximal muscle weakness. Although progress of the disease is similar to DMD it is slower. Patients on average become chair-bound at about 30 years of age and although some live to 70 years, the mean age of death is in the 40s.

Identification of the genetic lesion in Duchenne and Becker muscular dystrophies

Early studies on the enzymes and proteins of muscle from patients with DMD indicated that the isoform composition of these components of the tissue often resembled that of foetal tissue rather than that associated with normal muscle at a similar state of development. This lead to the hypothesis that the condition involved disturbance of the normal development process in muscle. It is now realised that these changes are secondary to the condition and merely reflect the regeneration of the muscle that is occurring in response to the dystrophy. An important observation was that made by Sibley & Lehninger (1949) who reported an increase of aldolase in the serum of patients with DMD. It soon became clear that the level of a range of soluble sarcoplasmic enzymes were elevated in the serum of patients with DMD and female carriers of the condition. In view of these findings the serum level of creatine phosphokinase, one of the most abundant enzymes in skeletal muscle, was eventually adopted as a diagnostic test for DMD homo and heterozygotes. These observations indicated that the muscle cell membrane had become more permeable to the soluble proteins of the cell and suggested that a membrane defect was a primary or secondary consequence of the lesion. It is considered that the increased membrane permeability results in raised intracellular calcium levels and thus stimulates the degenerative processes in the cell.

Early in the investigation of DMD it became clear that the disease is X-linked and recessive but it was not until the late 1970s and early 1980s that progress was made in localising the responsible gene on the X chromosome. At this time a number of females who exhibited X:autosomal translocations were described showing all the symptoms of DMD (for a review, see Boyde *et al.*, 1986). The condition was very rare and in each case the break-point on the X chromosome occurred in the dark Xp21 band of the short arm. It was concluded that the break disrupted the expression of the gene which was located in this region and that the normal allele was inactivated. Linkage studies on families with DMD using cloned DNA segments from the human X chromosome indicated that the segments most closely linked were those from the Xp21 region. Thus by the early 1980s it was concluded that the gene which had undergone mutation in DMD was located in the region Xp21. As the same gene is involved in Duchenne and Becker dystrophies these conditions are referred to as Xp21 myopathies or dystrophinopathies.

To obtain probes for the gene and hence identify it Worton and collaborators (Worton & Burgess, 1988) used DNA from a patient with the X:21 chromosome translocation to identify the breakpoint which would be expected to lie within the gene in question. An alternative approach that led to the identification and cloning of the gene was carried out by Kunkel and collaborators (see Kunkel, 1989). This group took advantage of a rare condition in which the Xp21 region of the chromosome was deleted. The patient with the deletion suffered from muscular dystrophy and other X-linked conditions including chronic granulomatous disease, retinitis pigmentosa and McLeod red cell phenotype. Normal DNA split into fragments by a restriction enzyme was mixed with 200 × excess DNA, from the patient, that had been broken into fragments by sheering stress. Hybrids of DNA containing such fragments will not ligate into plasmids. The mixture of the two samples of DNA was dissociated and reassociated by the phenol reassociation technique (pERT). Fragments of DNA corresponding to the missing region in the patient, which were the only fragments that would ligate into plasmids, were cloned and identified by their inability to hybridise with his DNA. These so-called pERT clones were screened for those that did not hybridise with DNA from patients with DMD. In this way regions of the gene were identified and expanded by chromosome walking. Using the probes so obtained regions of the gene conserved in a variety of mammalian species were identified. These probes enabled a low abundance 16 kilobase RNA to be detected in skeletal muscle. From this a cDNA library was prepared and the nucleotide sequence of the gene determined.

The dystrophin gene and its product

The dystrophin gene, named after the protein product that is absent or present in a modified form in Xp21 myopathies, is the largest gene so far identified. It spans 2300 kilobases and is 12 times the size of one of the largest genes known, that of coagulation factor VIII. It represents about 1% of the X chromosome and 0.05% of the entire human genome. This would appear to be an exceptionally high fraction in view of the fact that the human genome is estimated to control the synthesis of 50 000 to 100 000 proteins. To date, 79 exons constituting only 0.6% of the gene and giving rise to a cDNA of 14 kilobases have been identified. Introns vary in size from a few to 160–180 kilobases.

From the size of the cDNA it was concluded that the gene coded for a large protein of molecular weight about 400 000. In the earlier stages of the investigations, before detailed sequence information became available, attention was focused on nebulin, a known myofibrillar protein of similar size to the postulated gene product. It soon became clear that nebulin was not the normal product of the gene because if care were taken to prevent proteolysis it could be detected in dystrophic muscle. When the nucleotide sequence of the gene became known it was shown to correspond to a polypeptide chain of 3865 residues and a molecular weight of 427 kDa (Koenig *et al.*, 1988). This protein, the sequence of which did not correspond to any previously known molecule, was named dystrophin (Fig. 6.1). Some indication of the function of dystrophin is apparent from its multi-domain structure. The N-terminal domain of 240 residues shows homology to the N-terminal region of α-actinin, a protein that has been shown to bind actin in this domain (Mimura & Asano, 1986). The second and largest domain is formed by a succession of 25 presumed triple helical segments (the fourth of which is truncated) with loose homology between them and which show weak homology to similar repeats in α and β-spectrin. A third domain extending over 150 residues is cysteine-rich and is similar in sequence to the carboxy terminus of α-actinin that contains two calcium-binding sites. The corresponding sites on dystrophin are rather rudimentary and may not be functional. The carboxy terminal domain consisting of 420 residues does not possess homology to any known protein.

Immunocytochemical studies have now established the localisation of dystrophin in or close to the cell membrane. Early studies suggested that the protein is present in the T-tubule system but this has not been generally confirmed. It is probable that the staining observed in this investigation was caused by contamination of the preparations with cell membranes. The very small quantities of dystrophin present in muscle make it difficult to isolate amounts large enough to undertake a detailed study of its properties. The isolated protein seen by electron microscopy is a rod-like molecule with suggestions of globular ends. Estimates of the length of the molecules from the electron microscopic studies vary from 100–120 nm (Sato *et al.*, 1992) to 180 nm (Pons *et al.*, 1990). The linear molecule of diameter about 2 nm which appears to have globular ends has when isolated a tendency to form staggered side-to-side aggregates. These are usually dimers or tetramers and have a dumbbell structure (Sato *et al.*, 1992). The values for the length of the molecule obtained from studies on the isolated protein should be compared with the periodicity of 120 nm obtained by Cullen *et al.* (1990, 1991) using gold-labelled antibodies to segments of dystrophin represented by polypeptides composed of residues 1181–1388 and the last 17 C-terminal amino acids. From these studies the authors conclude that dystrophin forms a network structure with the rod-like region lying 15 nm from the cytoplasmic face of the membrane and with the C-terminus inserted in the membrane. No significant labelling of the T-tubules was observed in this study. On the basis of the proposed distribution it was calculated that dystrophin represented 0.0019% of the total muscle protein, a value similar to that estimated by Hoffman *et al.* (1987) from a comparison of the intensity of band staining with those of nebulin in Western blot analysis.

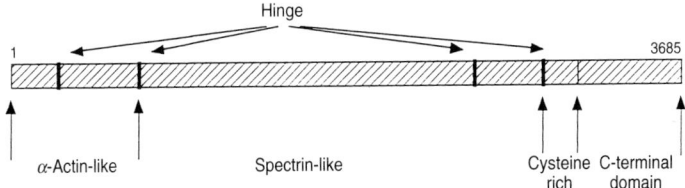

Fig. 6.1. Schematic representation of the dystrophin molecule indicating domains. Hinge regions are rich in proline and may represent segments that confer flexibility on the molecule (Koenig & Kunkel, 1990). Numbers indicate residue positions.

Isoforms of dystrophin

The dystrophin gene appears to be expressed in significant amounts only in muscle and neuronal tissues. From mRNA data it has been reported that smooth muscle contains 5–10% of the dystrophin present in skeletal muscle. This conclusion appears to be at variance with the report that chicken gizzard is a source of dystrophin in relatively high yield (Pons *et al.*, 1990). Neuronal tissue also contains less dystrophin than muscle (for a review, see Walsh *et al.*, 1989). As is the case with many other proteins dystrophin exists in a number of isoforms that are characteristic of the tissue from which they are derived. These are produced in the main by alternative splicing of the complementary RNA but in the case of the brain there appears to be alternative promoter and regulator sequences (Chelly *et al.*, 1990). In human and rat brain this results in three N-terminal amino acids replacing the 11 found in muscle dystrophin. Differences between the dystrophins of skeletal, cardiac and smooth muscles and brain are also produced by alternative splicing in the C-terminal region (Feener *et al.*, 1989).

Short forms of dystrophin

A number of proteins with epitopes identical to those of dystrophin have been identified using antibodies specific for the C-terminal region of the molecule and by characterising forms of mRNA present in cells. The best known of these have been designated DP 116, DP 71 and DP 45 with the number indicating the size in kDa. They are derived from the two C-terminal domains of the dystrophin gene (Fig. 6.2). In all cases most of the molecule is identical with the corresponding sequence in dystrophin but all have short peptide sequences at the N-termini that are characteristic of these molecules. In addition DP 71 also has a short unique C-terminal sequence. Much has to be learnt about the function and tissue specificity of these proteins for often they occur together with dystrophin. DP 116 appears

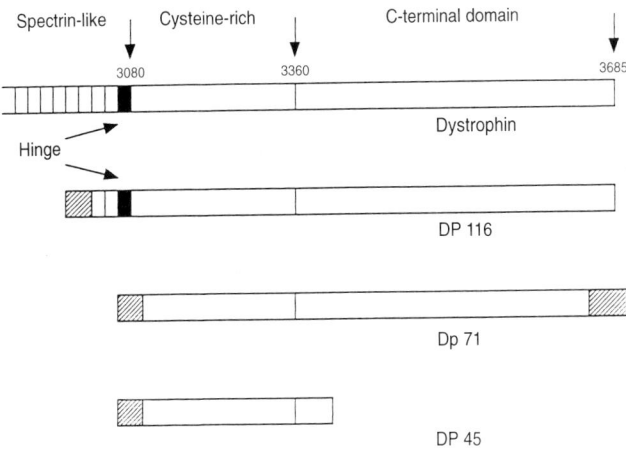

Fig. 6.2. Schematic diagrams indicating the relationship between dystrophin and the short forms of the protein. Hatched areas represent regions of the polypeptide sequence that are not identical to the corresponding regions in dystrophin (DP). Numbers indicate residue positions.

to be specific for peripheral nerve (for a review, see Fabbrizio *et al.*, 1994).

Utrophin

Soon after the discovery of dystrophin evidence was obtained for the presence of a protein similar in size to dystrophin that reacted with many of the antibodies specific for dystrophin but which was not identical with it. This protein, which was distinguished by the name dystrophin-related protein, DRP, or utrophin was shown to be of autosomal origin with the gene located on chromosome 6 (Love *et al.*, 1989, 1993). Its sequence is 85% identical with dystrophin but utrophin is not present in the sarcolemma of normal muscle where it is restricted to the myoneural junction. Its expression is upregulated in DMD and it is present in the sarcolemma of foetal and regenerating

156 kDa protein, which in this model is considered to bind to laminin, and the transmembrane protein 43 kDa are encoded by a single messenger RNA. It may be significant for the function of dystrophin that in DMD and in the *mdx* mouse the amount of the 156 kDa glycoprotein is reduced by 80–90% (Ervasti *et al.*, 1990). There is evidence that several of the DAGs interact with dystrophin, possibly in the cysteine-rich region of the C-terminal domains (Suzuki *et al.*, 1993).

The homology of the N-terminal region with α-actinin strongly suggests that actin binds to this region of dystrophin. By the use of proton NMR two N-terminal regions of dystrophin have been identified as sites to which actin binds (Levine *et al.*, 1992). Actin-binding site 1 is located at residues 17–26 and actin-binding site 2 is in the region of residues 128–156. The corresponding binding sites on actin are in the regions represented by residues 83–117 and residues 350–375, respectively. From a knowledge of the structure of the actin monomer and its proposed arrangement in F-actin these regions would be in the exposed domain of the actin filament.

A particular feature of actin-binding site 1 is the tetrapeptide Lys-Thr-Phe-Thr which is also found in the actin-binding regions of β-spectrin, the α-actinins and actin gelation factor. When the sequences of these proteins are aligned so that the tetrapeptides are coincident striking homologies can be seen between all the proteins (Fig. 6.4). In view of the known actin-binding properties of the other proteins this is excellent confirmatory evidence for the specific interaction of actin with dystrophin. The actin-binding site of dystrophin is not present in the N-terminal region of the spectrin-like domain as is the case with β-spectrin but at the N-terminus with homologies with α-actinin. Thus with regard to its actin-binding properties dystrophin can be considered as a 'stretched' spectrin.

The precise nature of the interaction between actin and dystrophin is probably quite complex for both proteins are organised in the cytoskeleton as polymers. If the analogy with spectrin is valid, dystrophin would be expected to exist as an antiparallel dimer with actin interacting at both ends of the dimer; indeed, electron microscopic studies of isolated dystrophin give some support to this view which is incorporated in the scheme (Fig. 6.3) proposed by Ervasti & Campbell (1991). The positions of the two sites on the actin molecule at which dystrophin binds are such as to suggest that they are independent and not parts of one interaction site. In this case the corresponding sites on dystrophin will also be independent. As in the cytoskeleton the actin filament is a double helical arrangement of linear aggregates of globular G-actin subunits the possibility exists that the dystrophin binds to two adjacent subunits of actin either in the same or in the parallel chain of the

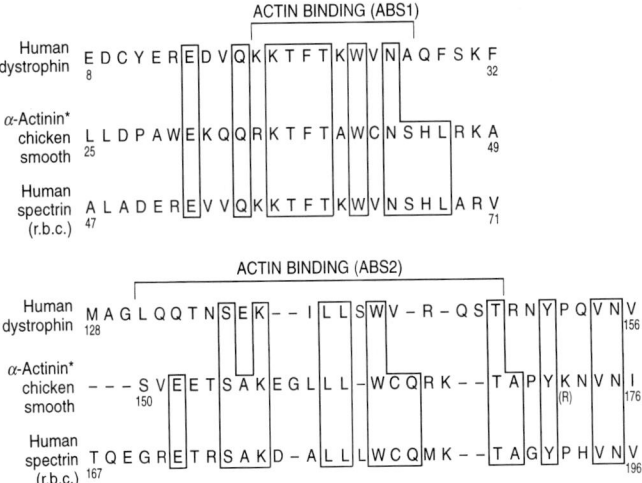

Fig. 6.4. Comparison of the amino acid sequences of the actin-binding regions of human dystrophin (ABS1 and ABS2) with the corresponding segments of α-actinin and β-spectrin. *The corresponding sequence of chicken skeletal α-actinin is identical except at residue 172 (Levine *et al.*, 1992). Numbers indicate residue positions. r.b.c., erythrocytes.

double helix. Much more work, however, is required to define in detail the nature of the interaction of actin and dystrophin.

By analogy with the role of spectrin in the erythrocytes dystrophin probably interacts with a number of other proteins in addition to actin and the dystrophin-associated proteins. Possible candidates include ankyrin, protein 4.1, protein 4.9, adducin, tropomyosin and the glycophorins.

The evidence that dystrophin is a component of the cytoskeletal membrane system is overwhelming but its precise role in membrane function is not understood. In the human its absence results in a leakiness of the muscle membrane that allows soluble sarcoplasmic enzymes to appear in the blood. Its absence may also be responsible for the increased intracellular calcium levels that lead to necrotic changes occurring in the tissue. There do, however, appear to be species differences in that in the *mdx* mouse the plasma creatine kinase levels fall to low values after a peak at about 3–4 weeks, despite the fact that dystrophin is absent from the muscle throughout life. Nevertheless, some modifications of the mechano-transducing channels of the myotubes have been reported. A higher proportion of the calcium permeable ionic channels of 7-day-old mice are stretch-inactivated in the *mdx* mouse than is the case with the normal control (Franco & Lansman, 1990). This work suggests that dystrophin may have a role in maintaining the integrity of the ion channels in muscle. Similar conclusions can be drawn from the report from the mature *mdx* mouse muscle fibre which, although it is relatively normal in its contractile behaviour, exhibits decreased osmotic stability (Menke & Jockusch, 1991) and elastic modulus of area expansion (Hutter *et al.*, 1991) compared with the wild-type control. Evidence in support of the view that the membrane integrity is changed is the fact that the absence of dystrophin leads to a loss of the dystrophin-related proteins, particularly the 156 kDa glycoprotein, from muscle fibres but not from the myoneural junction or the sarcolemma of intrafusal fibres (Matsumura *et al.*, 1992). This suggests that the dystrophin-deficient muscle fibres lack the normal interaction between the sarcolemma and the intracellar cytoskeleton. The importance of the dystrophin-associated glycoproteins for normal muscle cell function is indicated by the recent identification of the biochemical lesion in severe childhood autosomal recessive muscular dystrophy (SCARMD). This condition is a progressive muscular dystrophy prevalent in North Africa and shares many clinical features with Duchenne muscular dystrophy. Nevertheless, dystrophin and utrophin are expressed normally in this condition but there are reductions in the expression of the 50 and 35 kDa dystrophin-related glycoproteins (Matsumura *et al.*, 1992).

Gene changes in Xp21 myopathies

The availability of genomic and cDNA probes for the dystrophin gene and antibodies to dystrophin and dystrophin-related proteins has enabled Xp21 myopathies to be diagnosed with accuracy and in most cases the nature of the mutation defined. For example, Den Dunnen *et al.* (1989) were able to define precisely the actual mutation in 65% of the 194 cases of Xp21 myopathies studied (Fig. 6.5). This is probably a conservative figure for not all the probes available were used extensively in this study. The dystrophin gene exhibits virtually all the possible types of mutation, that is, point mutations, deletions, duplications, translocations, insertions and inversions. Translocations, insertions and inversions are particularly difficult to detect in intronic sequences. Deletions have been detected over the whole gene but their distribution, as well as that of the duplications, is not random. A major 'hot spot' for deletion breakpoints occurs close to the middle of the cDNA in the P20 intron region whereas a minor hot spot is located closer to the 5' end of the gene (Fig. 6.5). Most of the breakpoints occur in introns but because of the size of the deletions they can also lead to exon loss. The deletions vary in size from a few kilobases to 700 kilobases but surprisingly there is no correlation between their size and the severity of the resulting clinical condition. The largest deletion so far recorded represented 46% of the molecule and includes much of the spectrin domain. This occurred in a patient with BMD who lived until he was 61 years of age with a shortened dystrophin of molecular mass about 200 kDa in his muscles (England *et al.*, 1990).

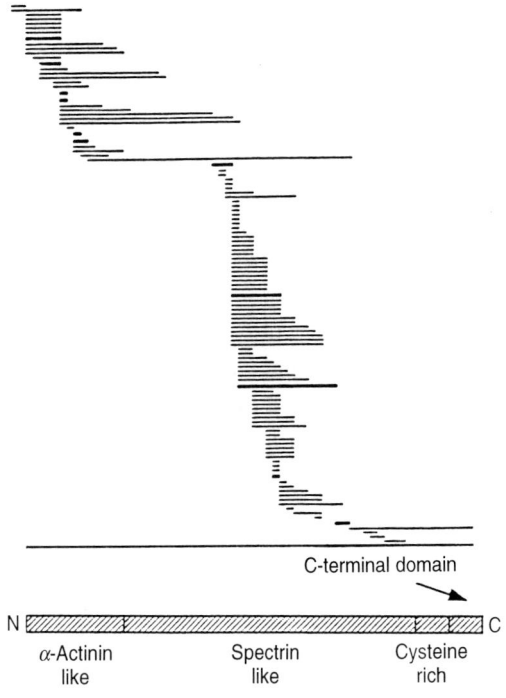

Fig. 6.5. Schematic representation of the mutations of the DMD gene observed in 194 cases using fragments of dystrophin cDNA as probes. Gene deletions are indicated by a thin line and duplications by a thick line.
Reproduced after Den Dunnen *et al.* (1989).

There is no obvious difference in the distribution of the deletions between patients with DMD and BMD that would explain the clear difference in the severity of the condition in these two groups. It has been proposed that in patients with BMD the deletion is such that the reading frame is not disrupted with the result that a truncated protein is synthesised (Monaco *et al.*, 1988). On the other hand, it is suggested that because the reading frame is disrupted in DMD the protein synthesised is unstable and is rapidly broken down. This would explain the fact that in DMD little or no dystrophin or its mutant form can be detected whereas in BMD muscle a band that usually, but not in all cases, has a lower molecular weight than dystrophin can be detected by antibody staining. Most cases fit in with this hypothesis but there do appear to be some exceptions.

Despite the advances that have been made showing a clear correlation between the mutant dystrophin gene and the Xp21 myopathies the precise nature of the lesions that are responsible for the muscle cell failure in DMD or BMD is not clear. There are suggestions that in the absence of dystrophin the membrane function is impaired because of a reduction in the amounts of DAGs present. Nevertheless, it appears that not all species have the same requirement for dystrophin and DAGs to maintain a functional muscle membrane. This fact and the recent demonstration of the widespread distribution of dystrophin-related proteins, the synthesis of which is under autosomal control, indicates that much is to be learnt about the role of the dystrophin family of proteins. It seems likely that mutation of the dystrophin gene may have consequences other than that resulting in the absence of dystrophin and associated proteins which are also important for normal membrane function (Ervasti *et al.*, 1990). In this context it may be significant that the deletion 'hot spots' occur in the larger introns. It might be expected that evolution would result in the loss of large introns that are prone to pathological deletions. Their persistence in the gene suggests that they may contain sequences essential for the organism and which have a function that is not yet understood. They may even contain genes that can be expressed independently of the dystrophin gene.

Animal models

Until the early 1980s the widely used animal model for muscular dystrophy research was the *dy* mouse. Although the progressive deterioration of muscle function was in many ways similar to DMD the inheritance of the condition is autosome linked and therefore unlikely to involve the same gene as the Xp21 myopathies. More recently the *mdx* mouse, which arose spontaneously in an inbred C57BL/10 strain (Bulfield *et al.*, 1984), has been widely used as an animal model for the Xp21 myopathies. This mutant mouse shows elevated serum creatine kinase as in DMD and is the consequence of a point mutation in one of the exons that leads to premature termination of the polypeptide chain (Sicinski *et al.*, 1990). The truncated protein is presumed to be unstable and is broken down rapidly. Neither dystrophin nor its truncated fragment can be detected in *mdx* mouse muscle. Nevertheless, it is difficult to detect marked differences in the physiological properties in the muscles of older animals when they are compared with controls (Dangain & Vrbova, 1984). At 3–4 weeks of age the muscles of *mdx* mice are clearly weaker than normal controls. As the animals grow older, however, the muscles recover and often become bigger and stronger than in controls of a similar age. Necrotic foci are seen in the *mdx* muscle and the number of cells with central nuclei increases reaching almost 100% in the older animals. As is the case with DMD the end-plate region of the *mdx* mouse muscle stains with antibody to dystrophin but this probably results from the dystrophin-related proteins localised in this region. One of the differences between DMD and *mdx* mouse muscle is the marked replacement of the muscle with fibrous tissue that occurs in the human conditions. It has been suggested that this is much less apparent in the *mdx* mouse because of the marked capacity of mouse tissue to regenerate. Such an explanation is supported by centralisation of the nuclei which is considered to be an index of regenerative changes in the muscle. It is not clear how the regeneration of mouse muscle enables it to dispense with dystrophin because there is no evidence that regenerating muscle has any less of requirement for the protein than the mature tissue. Further evidence of species differences in the requirement for dystrophin is presented by the dog and cat. The protein is absent from the muscle of a mutant form of the Golden Retriever in which the severity of the condition is similar to DMD (Cooper *et al.*, 1988). A dystrophin-deficient mutant of the cat has also recently been discovered which is characterised by hypertrophy of certain muscles and a relatively mild dystrophy (Carpenter *et al.*, 1989).

Treatment

An important aim of research into muscular dystrophy is to reduce the incidence of the mutant gene in a population to as low a level as is possible. In the case of the Xp21 myopathies this aim is attainable by genetic counselling now that specific gene probes are available to identify carriers and foetuses carrying the mutant gene. A satisfactory method of treatment of the condition is also required for there is a total of several million afflicted individuals in the world at the moment. There will be a continuing need for treatment even if the currently identified pool of mutant genes could be eliminated because of the high mutation rate that is present in DMD. For this reason the mutant gene pool would stabilise at about a third of its current value and might even rise if treatment is adequate to enable homozygotes to lead a normal life and to reproduce.

The principal methods that can be used to treat the Xp21 myopathies are pharmacological intervention, gene therapy and cell therapy. Of these approaches the first is the only one that requires a detailed understanding of the nature of the lesion to develop an effective pharmacological agent in a logical way. The present state of knowledge does not offer much guidance as to the rationale for the development of such a drug. Observation strongly suggests that the lesion is a membrane defect which offers some promise that it might be possible to intervene more effectively than if the site of the lesion were deep inside the cell. In this respect, however, it should be mentioned that attempts to treat DMD with calcium antagonists have been made to circumvent the increased permeability to that cation. They were discontinued because of the widespread deleterious effects of these reagents. An understanding of the manner in which the *mdx* mouse modulates its metabolism to compensate for the absence of dystrophin might give a valuable lead for the development of an effective therapy.

Ideally the treatment of choice would be to replace the defective gene with the normal dystrophin gene. To ensure that the gene is expressed in all tissues in which dystrophin is normally present it would need to be introduced into the oocyte. There has been some success in applying this technique to animals but it has not yet been applied to humans. There are considerable technical difficulties in transferring and expressing into a somatic cell a gene as large and as complex as the dystrophin gene. Nevertheless, this has been achieved with transgenic mice where the expression of dystrophin eliminated the morphological and immunohistochemical symptoms of muscular dystrophy (Cox *et al.*, 1993). At the present time a promising approach is to introduce the gene directly into the muscle. Direct *in-vivo* transfer of DNA into postnatal tissue has been achieved using a variety of methods. These include DNA encapsulated in liposomes containing viral envelope receptor proteins, coprecipitated with calcium phosphate, coupled to polylysine glycoprotein carrier complex, associated with cationic lipid vesicles, etc.

Wolf *et al.* (1990) have reported the successful expression of reporter molecules after the injection of relatively simple DNA and RNA constructs into mouse muscle. For example, firefly luciferase could be detected in muscle 60 days after injection. The reporter molecule could be detected up to 400 nm from the site of injection but most of the DNA was not integrated in the nucleus but persisted in a circular non-replicating form. The mechanism of entry is not known but the levels of reporter molecules introduced in this way were significantly higher in muscle than in other tissues. It is possible that the structure of muscle with its many transverse tubules facilitates penetration of the DNA. Another possibility is that the nucleotide penetrates cells that are damaged by the injection and then recover.

Cellular therapy is the one method of direct introduction of the dystrophin gene into muscle that has undergone clinical trials. It is based on the observation made some time ago that when myoblasts are injected into the muscle of an animal they can fuse with the host stem cells and form myotubes and ultimately mature muscle fibres expressing proteins that are products of the host and donor genes (Watt *et al.*, 1984). When this experiment was repeated by injecting myoblasts from a normal control into muscle of the *mdx* mouse a proportion of the fibres were shown to contain dystrophin. This was located in the membrane although sometimes the distribution was more patchy than is normally the case. To overcome the problem of rejection in this investigation the host mice were the *nu/nu* strain with a deficient immune system that will accept grafts from other strains (Partridge *et al.*, 1989). In the animal experiments it was not possible to determine whether the muscle properties had been improved by restoring dystrophin to the fibres, but the clinical application of this method of restoring dystrophin to the muscles of patients with DMD has aroused much interest. A number of problems exist if myoblast transfer is to be used as a treatment for DMD. Large numbers of myoblasts, usually about a million cells, are required per injection. The sites of injection of the myoblasts must be close together because the effects in the animal experiments only extended for about ten fibres from the site. Much has yet to be learnt about the immune reactions produced by myoblast transfer. For this reason the limited trials that have been carried out were conducted in the presence of immunosuppressive agents. Even if the procedure were successful in a single muscle it would be an enormous if not impracticable operation to inject myoblasts into all the muscles of the human body (Perry, 1992). There is also the problem of introducing dystrophin into the nervous system where it is also expressed and its absence may be responsible for the moderate cognitive impairment associated with DMD (Lidov *et al.*, 1990).

In recent years the approach to treatment has become more complicated with the discovery of the complex nature of the dystrophin family of proteins, their widespread distribution in tissues and their close association with a number of other proteins which appear to be essential for normal membrane function. It is very puzzling that species vary in the effect of the absence of dystrophin on muscle function. This might be explained by the possibility that autosomal utrophin, under different genetic control, may in some cases be substituted for dystrophin with which it has a marked homology. In that case a more rewarding approach to treatment would be to attempt to upgrade the expression of the utrophin gene rather than attempt to introduce the dystrophin gene into tissues. Likewise, a similar approach could apply to the dystrophin-associated proteins, the amounts of which in DMD and *mdx* mouse muscles, are depressed in the absence of dystrophin. Nevertheless, there is no evidence that their genes are abnormal in the dystrophic organism. Experience with SCARMD clearly indicates that some of these proteins at least are vital for muscle function irrespective of the presence of dystrophin. The recent advances in our knowledge of the nature of the lesion in DMD have increased our understanding of the cytoskeletal–membrane system of muscle and other tissues. They have indicated the shortcomings of existing knowledge but it is clear that when the function of dystrophin is finally elucidated new light will be thrown on the nature of the cytoskeletal–membrane systems of tissues generally.

References

Chapter 1

Armitage, J. P. (1992). Bacterial motility and chemotaxis. *Science Progress Oxford*, **76**, 451–77.

Berliner, E., Young, E. C., Anderson, K., Mahtani, H. & Gelles, J. (1995). Failure of single-headed kinesin to track parallel to microtubule protofilaments. *Nature*, **373**, 718–21.

Cheney, R. E. & Mooseker, M. A. (1992). Unconventional myosins. *Current Opinion in Cell Biology*, **4**, 27–35.

Collins, J. H. (1991). Myosin light chains and troponin C: structural and evolutionary relationships revealed by amino acid sequences. *Journal of Muscle Research and Cell Motility*, **12**, 3–25.

Endow, S. A. (1991). The emerging kinesin family of microtubule motor proteins. *Trends in Biochemical Science*, **16**, 221–5.

Euteneuer, U., Koonce, M. P., Pfister, K. K. & Schliwa, M. (1988). An ATPase with properties expected for an organelle motor of the giant amoeba *Reticulomyxa*. *Nature*, **332**, 176–8.

Goodson, H. V. & Spudich, J. A. (1993). Molecular evolution of the myosin family. Relationship derived from amino acid sequences. *Proceedings of the National Academy of Sciences, USA*, **90**, 659–63.

Hammer, J. A. (1994). The structure and function of unconventional myosins: a review. *Journal of Muscle Research and Cell Motility*, **15**, 1–10.

Hinshaw, J. E. & Schmid, S. L. (1995). Dynamin self-assembles into rings suggesting a mechanism for coated vesicle budding. *Nature*, **374**, 190–2.

Hirose, K., Lockhart, A., Cross, R. A. & Amos, L. A. (1995). Nucleotide-dependent angular change in kinesin motor domain bound to tubulin. *Nature*, **376**, 277–9.

Hoenger, A., Sablin, E. P., Vale, R. D., Fletterick, R. J. & Milligan, R. A. (1995). Three-dimensional structure of tubulin-motor protein complex. *Nature*, **376**, 271–4.

Magari, Y., Sugiyama, S., Muramoto, K., Maekawa, Y., Kawagishi, I., Imae, Y. & Kudo, S. (1994). Very fast flagellar rotation. *Nature*, **371**, 752.

Mooseker, M. S. & Cheney, R. E. (1995). Unconventional myosins. *Annual Review of Cell Developmental Biology*, **11**, 633–75.

Nishizaka, T., Yagi, T., Tanaka, Y. & Ishiwata, S. (1993). Right-handed rotation of an actin filament in an *in-vitro* motile system. *Nature*, **361**, 269–71.

Pollard, T. D. & Korn, E.D. (1973). *Acanthamoeba* myosin. 1. Isolation from *Acanthamoeba castellani* of an enzyme similar to muscle myosin. *Journal of Biological Chemistry*, **248**, 4682–90.

Romberg, L. & Vale, R. D. (1993). Chemomechanical cycle of kinesin differs from that of myosin. *Nature*, **361**, 168–70.

Schnapp, B. J. (1995). Two heads are better than one. *Nature*, **373**, 655–6.

Shpetner, H. S. & Vallee, R. B. (1992). Dynamin is a GTPase stimulated to high levels of activity by microtubules. *Nature*, **355**, 733–5.

Snyder, J. A. & McIntosh, J. R. (1976). Biochemistry and physiology of microtubules. *Annual Review of Biochemistry*, **45**, 699–720.

Spudich, J. A. (1994). How molecular motors work. *Nature*, **372**, 515–18.

Squire, J. (1983). Molecular mechanisms in muscular contraction. *Trends in Neuroscience*, **6**, 409–13.

Taylor, E. W. (1993). Variations on a theme of movement. *Nature*, **361**, 115–16.

Vale, R. D. (1992). Microtubule motors: many new models off the assembly line. *Trends in Biochemical Sciences*, **17**, 300–4.

Walker, R. A. & Sheetz, M. P. (1993). Cytoplasmic microtubule-associated motors. *Annual Review of Biochemistry*, **62**, 429–51.

Chapter 2

Barany, M. (1967). ATPase activity of myosin correlated with speed of muscle shortening. *Journal of General Physiology*, **50**, 197–218.

Burton, K. (1992). Myosin step size: estimates from motility assays and shortening muscle. *Journal of Muscle Research and Cell Motility*, **13**, 590–607.

Finer, J. T., Simmons, R. M. & Spudich, J. A. (1994). Single myosin molecule mechanics: piconewton force and nanometre steps. *Nature*, **368**, 113–19.

Geeves, M. A. (1991). The dynamics of actin and myosin interaction and the cross-bridge model of muscle contraction. *Biochemical Journal*, **274**, 1–14.

Haselgrove, J. C. (1983). Structure of vertebrate muscle. In: *Handbook of Physiology*, Section 10, *Skeletal Muscle*, ed. by L. D. Peachey, R. H. Adrian & G. R. Geiger. pp. 143–71. American Physiological Society, Bethesda, MD, USA.

Holmes, K. C., Popp, D., Gebhard, W. & Kabsch, W. (1990). Atomic model of the actin filament. *Nature*, **347**, 44–9.

Huxley, H. E. (1990). Sliding filaments and molecular motile systems. *Journal of Biological Chemistry*, **265**, 8347–50.

Huxley, H. E. & Kress, M. (1985). Cross-bridge behaviour during muscle contraction. *Journal of Muscle Research and Cell Motility*, **6**, 153–61.

Irving, M., Allen, T. S.-C., Sabido-David, C., Craik, J. S., Brandmeler, B., Kendrick-Jones, J., Corrie, J. E. T., Trentham, D. R. & Goldman, Y. E. (1995). Tilting of the light-chain region during step length changes and active force generation in skeletal muscle. *Nature*, **375**, 688–91.

Ishijima, A. & Yanagida (1995). Direct measurements of unitary forces and steps of single myosin molecules. In: *Calcium as Cell Signal*, ed. by K. Maruyama, Y. Nonomura & K. Kohama, pp. 382–8. Tokyo, New York: Igahu-Shoin.

Jontes, J. D., Wilson-Kubalek, E. M. & Milligan, R. A. (1995). A 32° tail swing in brush border myosin I on ADP release. *Nature*, **378**, 751–3.

Kabsch, W., Mannherz, H. G., Suck, D., Pai, E. F. & Holmes, K. C. (1990). Atomic structure of actin:DNAse complex. *Nature*, **347**, 37–44.

Kishino, A. & Yanagida, T. (1988). Force measurements by micromanipulation of single actin filament by glass needles. *Nature*, **334**, 74–6.

Levine, B. A., Moir, A. J. G., Trayer, I. P. & Williams, R. J. P. (1990). Nuclear magnetic resonance studies of calcium-modulated proteins and actin-myosin interaction. In: *Molecular Mechanisms of Muscle Contraction*, ed. by J. Squire, pp. 171–209. London: Macmillan.

Lowey, S., Waller, G. S. & Trybus, K. M. (1993). Skeletal muscle light chains are essential for physiological speeds of shortening. *Nature*, **365**, 454–6.

Lymn, R. W. & Taylor, E. W. (1971). Mechanism of adenosine triphosphate hydrolysis by actomyosin. *Biochemistry*, **10**, 4617–24.

Molloy, J. E., Burns, J. E., Kendrick-Jones, J. E., Tregear, R. T. & White, D. C. S. (1995). Movement and force produced by a single myosin head. *Nature*, **378**, 209–12.

Mornet, D., Bonet, A., Audemarde, E. & Bonicel, J. (1989). Functional sequences of the myosin head. *Journal of Muscle Research and Cell Motility*, **10**, 10–24.

Mueller, H. & Perry, S. V. (1962). Degradation of heavy meromyosin by trypsin. *Biochemical Journal*, **85**, 431–9.

Oplakta, A. & Tirosh, R. (1973). Active streaming in actomyosin solutions. *Biochimica Biophysica Acta*, **305**, 684–8.

Otsuki, I., Maruyama, K. & Ebashi, S. (1986). Regulatory and cytoskeletal proteins of vertebrate skeletal muscle. *Advances in Protein Chemistry*, **38**, 1–67.

Perry, S. V. (1989). The interaction of actin and myosin 40 years ago. *Biochimica Biophysica Acta*, **1000**, 159–62.

Perry, S. V. & Cotterill, J. (1965). Interaction of actin and myosin. *Nature*, **206**, 161–3.

Rayment, I., Holden, H. M., Whittaker, M., Yohn, C. B., Lorenz, M., Holmes, K. C. & Milligan, R. A. (1993b). Structure of the actin myosin complex and its implications for muscle contraction. *Science*, **261**, 58–65.

Rayment, I., Rypniewski, W. R., Schmidt-Base, K., Smith, R., Tomchick, D. R., Benning, M. M., Winkelmann, D. A., Wesenberg, G. & Holden, H.M. (1993a). Three-dimensional structure of myosin subfragment-1: a molecular motor. *Science*, **261**, 50–8.

Spudich, J. A. (1994). How molecular motors work. *Nature*, **372**, 515–18.

Squire, J. (1988). Invisible actin makes its debut. *Nature*, **335**, 590–1.

Toyoshima, Y. Y., Kron, S. J., McNally, E. M., Niebling, K. R., Toyoshima, C. & Spudich, J. A. (1987). Myosin subfragment-1 is sufficient to move actin filaments *in vitro*. *Nature*, **328**, 536–9.

Toyoshima, Y. Y., Toyoshima, C. & Spudich, J. A. (1989). Bidirectional movement of actin filaments along tracks of myosin heads. *Nature*, **341**, 154–6.

Vibert, P. & Cohen, C. (1988). Domains, motions and regulation of the myosin head. *Journal of Muscle Research and Cell Motility*, **9**, 296–305.

Wegner, A. (1985). Sublety of actin assembly. *Nature*, **313**, 97–8.

White, H. D. & Rayment, I. (1993). Kinetic characterisation of reductively methylated myosin subfragment-1. *Biochemistry*, **32**, 9859–65.

Whittaker, M., Wilson-Kubalek, E. M., Smith, J. E., Faust, L., Milligan, R. A. & Sweeney, H. L. (1995). A 32 Å movement of smooth muscle myosin on ADP release. *Nature*, **378**, 748–51.

Woledge, R. C. (1988). Forced dilemma to come in muscular contraction. *Nature*, **334**, 655.

Yanadiga, T., Arata, Y. & Oosowa, F. (1995). Sliding distance of actin filament induced by myosin cross bridge during one ATP hydrolysis cycle. *Nature*, **316**, 366–9.

Yount, R. G. (1993). Subfragment 1: the first crystalline motor. *Journal of Muscle Research and Cell Motility*, **14**, 547–51.

Yount, R. G., Cremo, C. R., Crammer, J. C. & Kerwin, B. A. (1992). Photochemical mapping of the active site of myosin. *Philosophical Transactions of the Royal Society*, **336**, 55–61.

Xie, X., Harrison, D. H., Schlichting, I., Sweet, R. M., Kalabokis, V. N., Szent-Gyorgyi, A. G. & Cohen, C. (1993). Structure of the regulatory domain of scallop myosin at 2.8 Å resolution. *Nature*, **368**, 306–12.

Chapter 3

Ashley, C. C., Mulligan, I. P. & Lea, T. J. (1991). Ca^{2+} and activation mechanisms in skeletal muscle. *Quarterly Reviews of Biophysics*, **24**, 1–73.

Block, B. A., Imagawa, T., Campbell, K. P. & Franzini-Armstrong, C. (1988). Structural evidence for direct interaction between the molecular components of the transverse tubule/SR junction in skeletal muscle. *Journal of Cell Biology*, **107**, 2587–600.

Campbell, K. P., Leung, A. T. & Sharp, A. H. (1988). The biochemistry and molecular biology of the dihydropyridine-sensitive calcium channel. *Trends in Neurosciences*, **11**, 425–30.

Da Silva, A. C. R. & Reinach, F. C. (1991). Calcium binding induces conformational changes in muscle regulatory proteins. *Trends in Biochemical Science*, **16**, 53–7.

Fleischer, S. & Inui, M. (1989). Biochemistry and biophysics of excitation–contraction coupling. *Annual Review of Biophysics and Biophysical Chemistry*, **18**, 333–64.

Fliegel, L., Ohnishi, M., Carpenter, M. R., Khanna, V. K., Reithmeier, R. A. F. & MacLennan, D. H. (1987). Amino acid sequence of rabbit fast twitch muscle calsequestrin deduced from cDNA and peptide sequencing. *Proceedings of the National Academy of Sciences, USA*, **84**, 1167–71.

Ford, L. E. & Podolsky, R. J. (1970). Regenerative calcium release within muscle cells. *Science*, **167**, 58–9.

Francini-Armstrong, C. (1975). Membrane particles and transmission at the triad. *Federation Proceedings*, **34**, 1382–9.

Gillis, J. M. & Gerday, Ch. (1977). Calcium movement between myofibrils, parvalbumin and sarcoplasmic reticulum in muscle. In: *Calcium Binding Proteins and Calcium Function*, ed. by R. H. Wasserman, R. A. Corradino, E. Carafoli, R. H. Kretzinger, D. H. MacLennan & F. L. Seigel, pp. 193–6. Amsterdam: Elsevier/North Holland.

Grabarek, Z., Tan, R-Y. Wang, J., Tao, T. & Gergely, J. (1990). Inhibition of mutant troponin C activity by an intra-domain disulphide bond. *Nature*, **345**, 132–5.

Grabarek, Z., Tao, T. & Gergely, J. (1992). Molecular mechanism of troponin C function. *Journal of Muscle Research and Cell Motility*, **13**, 383–93.

Griffiths, P. J., Duchateau, J. J., Maeda, Y., Potter, J. D. & Ashley, C. C. (1990). Mechanical characteristics of skinned intact muscle fibres from the giant barnacle, *B. nubilis*. *Pflügers Archives*, **415**, 554–65.

Heizmann, C. W. (1984). Parvalbumin, an intracellular calcium-binding protein: distribution, properties and possible roles in mammalian cells. *Experentia*, **40**, 910–21.

Herzberg, O. & James, M. N. G. (1988). Refined crystal structure of troponin C from turkey skeletal muscle at 2.0 Å resolution. *Journal of Molecular Biology*, **203**, 761–79.

Herzberg, O., Moult, J. & James, M. N. G. (1986). A model for the Ca^{2+}-induced conformational transition of troponin C. *Journal of Biological Chemistry*, **261**, 2638–44.

Kretzinger, R. H. & Nockolds, C. E. (1973). Carp muscle calcium binding protein. *Journal of Biological Chemistry*, **248**, 3313–26.

Lai, F. A. & Meissner, G. (1989). The muscle ryanodine receptor and its intrinsic Ca^{2+} channel activity. *Journal of Bioenergetics and Biomembranes*, **21**, 227–46.

Lehman, W., Kendrick-Jones, J. & Szent-Gyorgyi, J. (1972). Myosin-linked regulatory systems: comparative studies. *Cold Spring Harbor Symposia on Quantitative Biology*, **37**, 319–30.

MacLennan, D. H. (1990). Molecular tools to elucidate problems in excitation–contraction coupling. *Biophysical Journal*, **58**, 1355–65.

MacLennan, D. H., Brandl, C. L., Korczak, B. & Green, N. M. (1985). Amino acid sequence of a $Ca^{2+} + Mg^{2+}$-dependent ATPase from rabbit muscle sarcoplasmic reticulum, deduced from its complementary DNA sequence. *Nature*, **316**, 696–700.

Martonosi, A. N. (1986). Regulation of calcium by the sarcoplasmic reticulum. In: *Myology*, 1st edn, vol. 1, ed. by A. G. Engel & B. Q. Banker, pp. 521–62. New York: McGraw-Hill Book Company.

Mitchell, R. D., Zimmerman, H. B. K. & Jones, L. R. (1988). Ca^{2+} binding effects on protein conformation and protein interactions of canine cardiac calsequestrin. *Journal of Biological Chemistry*, **263**, 1376–81.

Perry, S. V. (1994). Activation of the contractile mechanism by calcium. In: *Myology*, 2nd edn, vol. 1, ed. by A. G. Engel & C. Franzini-Armstrong, pp. 529–52. New York: McGraw-Hill Book Company.

Satyshur, K. A., Roa, S. T., Pyzalska, D., Drendel, W., Creaser, M. & Sundaralingam, M. (1988). Refined structure of chicken skeletal muscle troponin C in the two calcium state at 2 Å resolution. *Journal of Biological Chemistry*, **263**, 1628–47.

Strynadka, N. C. J. & James, M. N. G. (1989). Crystal structures of the helix-loop-helix calcium binding proteins. *Annual Review of Biochemistry*, **58**, 951–98.

Sweeney, H. L. & Stull, J. T. (1990). Alteration of cross-bridge kinetics by myosin light chain phosphorylation in rabbit skeletal muscle. Implications for regulation of actin–myosin interaction. *Proceedings of the National Academy of Science, USA*, **87**, 414–18.

Tada, M. & Kadama, M. (1989). Regulation of the Ca^+ pump ATPase by cAMP-dependent phosphorylation of phospholamban. *Bioessays*, **10**, 157–63.

Tsien, R. Y. (1988). Fluorescent measurement and photochemical manipulation of cytosolic free calcium. *Trends in Neurosciences*, **11**, 419–24.

Walker, J. W., Somlyo, A. V., Goldman, Y. E., Somlyo, A. P. & Trentham, D. R. (1987). Kinetics of smooth and skeletal muscle activation by laser pulse photolysis of caged inositol 1,4,5, trisphosphate. *Nature*, **327**, 249–52.

Chapter 4

Bailey, K. (1946). Tropomyosin: a new asymmetric protein component of muscle. *Nature*, **157**, 368.

Breitbart, R. E., Nguyen, H. T., Melford, R. M., Destree, A. T., Mahdavi, V. & Nadal-Ginard, B. (1988). Intricate combinatorial patterns of exon splicing generate multiple regulated troponin T isoforms from a single gene. *Cell*, **41**, 67–82.

Chalovitch, J. M., Chock, P. R. & Eisenberg, E. (1981). Mechanism of action of troponin–tropomyosin. *Journal of Biological Chemistry*, **256**, 575–8.

Dalgarno, D. C., Grand, R. J. A., Levine, B. A., Moir, A. J. G., Scott, G. M. M. & Perry, S. V. (1982). Interaction between troponin I and troponin C. *FEBS Letters*, **150**, 54–8.

Ebashi, S. (1963). Third component participating in the superprecipitation of actomyosin, *Nature*, **200**, 1010.

Grand, R. J. A., Henry, D. G., Moir, A. J. G., Perry, S. V., Trayer, I. P., Dalgarno, D. C., Levine, B. A. & Parker, S. B. (1983). Modulation by troponin C of the troponin I inhibition of skeletal actomyosin interaction. A PMR study. In: *Calcium Binding Proteins*, ed. by B. de Barnard, G. L. Sottocasa, G. Sandri, E. Carafoli, A. N. Taylor, T. C. Vanaman & R. J. P. Williams, pp. 379–80. Amsterdam: Elsevier/North Holland.

Haselgrove, J. C. (1972). X-ray evidence for a conformational change in the actin containing filaments of vertebrate striated muscle. *Cold Spring Harbour Symposia on Quantitative Biology*, **37**, 341–52.

Haselgrove, J. C. (1983). Structure of vertebrate striated muscle as determined by X-ray diffraction methods. *Handbook of Physiology*, Section 10, *Skeletal Muscle*, ed. by J. D. Peachey, R. H. Adrian & G. R. Geiger, pp. 143–71. Bethesda: American Physiological Society.

Heeley, D. H., Dhoot, G. K. & Perry, S. V. (1985). Factors determining the subunit structure of tropomyosin in skeletal muscle. *Biochemical Journal*, **226**, 461–8.

Heeley, D. H., Krystyna, G. & Smillie, L. B. (1987). The effects of troponin T fragments T1 and T2 on the binding of non-polymerisable tropomyosin to F-actin in the presence and absence of troponin I and troponin C. *Journal of Biological Chemistry*, **262**, 9971–8.

Holmes, K. C., Popp, D., Gebhard, W. & Kabsch, W. (1990). Atomic model of the actin filament. *Nature*, **347**, 44–9.

Huxley, H.E. (1972). Structural changes in actin and myosin-containing filaments during contraction. *Cold Spring Harbour Symposia on Quantitative Biology*, **37**, 361–78.

Ishikawa, T. & Wakabayashi, T. (1994). Calcium induced change in the three-dimensional structure of thin filaments of rabbit skeletal muscle as revealed by cryoelectron microscopy. *Biochemical and Biophysical Research Communications*, **203**, 951–8; **205**, 976.

Jackson, P., Amphlett, G. W. & Perry, S. V. (1975). The primary structure of troponin T and the interaction with tropomyosin. *Biochemical Journal*, **151**, 85–97.

Kobayashi, T., Tao, T., Gergely, J. & Collins, J. H. (1994). Structure of the troponin complex. Implications of photocross-linking of troponin I to troponin C thiol mutants. *Journal of Biological Chemistry*, **269**, 5725–9.

Krudy, G. A., Kleerekoper, Q., Guo, X. D., Howarth, J. W., Solaro, R. J. & Rosevear, P. R. (1994). NMR studies delineating spatial relationships within the cardiac troponin I–troponin C complex. *Journal of Biological Chemistry*, **269**, 23731–5.

Kumagai, H., Ebashi, S. & Takeda, F. (1955). Essential relaxing factor in muscle other than myokinase and creatine phosphokinase. *Nature*, **176**, 166.

Lehman, W., Craig, R. & Vibert, P. (1994). Ca^{2+}-induced tropomyosin movement in *Limulus* thin filaments revealed by three-dimensional construction. *Nature*, **368**, 65–7.

Levine, B. A., Moir, A. J. G. & Perry, S. V. (1988). The interaction of troponin I with the N-terminal region of actin. *European Journal of Biochemistry*, **172**, 389–97.

Lowy, J. & Vibert, P. J. (1972). Studies of the low angle X-ray pattern of molluscan smooth muscle during tonic contraction and rigor. *Cold Spring Harbour Symposia on Quantitative Biology*, **37**, 353–9.

Marsh, B. B. (1952). The effects of adenosine triphosphate on the fibre volume of a muscle homogenate. *Biochimica Biophysica Acta*, **9**, 247–60.

Pearlstone, J. R. & Smillie, L. B. (1981). Identification of a second binding region on rabbit skeletal troponin T for α-tropomyosin. *FEBS Letters*, **128**, 119–22.

Perry, S. V. (1994). Activation of the contractile mechanism by calcium. In: *Myology*, 2nd edn, vol. 1, ed. by A. G. Engel & C. Franzini-Armstrong, pp. 529–52. New York: McGraw-Hill Book Company.

Perry, S. V. & Grey, T. C. (1956*a*). A study of the effects of substrate concentration and certain relaxing factors on the magnesium activated myofibrillar ATPase. *Biochemical Journal*, **64**, 184–92.

Perry, S. V. & Grey, T. C. (1956b). Ethylenediaminetetraacetate and the ATPase activity of actomyosin systems. *Biochemical Journal*, **64**, 5.

Schachat, F. H., Diamond, M. S. & Brandt, P. W. (1987). Effects of different troponin T–tropomyosin combinations on thin filament activation. *Journal of Molecular Biology*, **198**, 551–4.

Smillie, L. B. (1979). Structure and function of tropomyosins from muscle and non-muscle sources. *Trends in Biochemical Science*, **4**, 151–4.

Syska, H., Wilkinson, J. M., Grand, R. J. A. & Perry, S. V. (1976). The relationship between biological activity and primary structure of troponin I from white skeletal muscle of the rabbit. *Biochemical Journal*, **153**, 375–87.

Szent-Gyorgyi, A. (1945). Studies on muscle. *Acta Physiologica Scandinavica*, **9** (suppl. 25), 3–115.

Talbot, J. A. & Hodges, R. S. (1981). Synthetic studies on the inhibitory region of rabbit skeleton troponin I. *Journal of Biological Chemistry*, **256**, 2798–802.

Weber, A. (1959). On the role of calcium in the activity of adenosine triphosphate hydrolysis by actomyosin. *Journal of Biological Chemistry*, **234**, 2764–9.

Chapter 5

Brenner, B. (1988). Effect of Ca^{2+} on cross-bridge turnover kinetics in skinned rabbit single psoas fibres: implications for regulation of muscle contraction. *Proceedings of the National Academy of Sciences, USA*, **85**, 3265–9.

Close, R. & Hoh, J. F. Y. (1969). Post-tetanic potentiation of twitch contractions of cross innervated rat fast and slow muscles. *Nature*, **221**, 179–81.

DeBelle, I. & Mak, A. S. (1987). Isolation and characterisation of tropomyosin kinase from chicken embryo. *Biochimica Biophysica Acta*, **925**, 17–26.

Grand, R. J. A., Wilkinson, J. M. & Mole, L. E. (1976). The amino acid sequence of rabbit cardiac troponin I. *Biochemical Journal*, **159**, 633–41.

Gusev, N. B., Dobrovolskii, A. B. & Severin, S. E. (1980). Isolation and some properties of troponin T kinase from rabbit skeletal muscle. *Biochemical Journal*, **189**, 219–26.

Heeley, D. H. (1994). Investigation of the effects of phosphorylation of rabbit striated muscle α-tropomyosin and rabbit skeletal muscle troponin T. *European Journal of Biochemistry*, **221**, 129–37.

Heeley, D. H., Moir, A. J. G. & Perry, S. V. (1982). Phosphorylation of tropomyosin and development in mammalian striated muscle. *FEBS Letters*, **146**, 115–18.

Heeley, D. H., Watson, M. H., Mak, A. S., Dubord, P. & Smillie, L. S. (1989). Effect of phosphorylation on the interaction and functional properties of rabbit striated muscle α-tropomyosin. *Journal of Biological Chemistry*, **264**, 2424–30.

Kendrick-Jones, J., Lehmann, W. & Szent-Gyorgyi, A. G. (1970). Regulation in molluscan muscles. *Journal of Molecular Biology*, **54**, 313–26.

Knighton, D. R., Pearson, R. B., Sowadski, J. M., Means, A. R., Eyck, L. F. T., Taylor, S. S. & Kemp, B. E. (1992). Structural basis of the intrasteric regulation of the myosin light chain kinases. *Science*, **258**, 130–5.

Lehmann, W. (1978). Thick filament-linked calcium regulation in vertebrate striated muscle. *Nature*, **274**, 80–1.

Manning, D. R. & Stull, J. T. (1979). Myosin light chain phosphorylation and phosphorylase A activity in rat extensor digitorum longus muscle. *Biochemical and Biophysical Research Communications*, **90**, 164–70.

Mittmann, K., Jaquet, K. & Heilmeyer Jr., L. M. G. (1990). A common motif of two adjacent phosphoserines in bovine, rabbit and human cardiac troponin I. *FEBS Letters*, **273**, 41–5.

Mittmann, K., Jaquet, K. & Heilmeyer Jr. L. M. G. (1992). Ordered phosphorylation of a duplicated minimal recognition motif for cAMP-dependent protein kinase present in cardiac troponin I. *FEBS Letters*, **302**, 133–7.

Moir, A. J. G., Cole, H. A. & Perry, S. V. (1977). The phosphorylation sites of troponin T from white skeletal muscle and the effects of interaction with troponin C on their phosphorylation by phosphorylase kinase. *Biochemical Journal*, **161**, 371–82.

Moir, A. J. G. & Perry, S. V. (1977). The sites of phosphorylation of rabbit cardiac troponin I by adenosine cyclic 3′:5′ monophosphate dependent protein kinase. Effect of interaction with troponin C. *Biochemical Journal*, **167**, 133–43.

Noland, T. A. & Kuo, J. F. (1993). Protein C kinase phosphorylation of cardiac troponin I and troponin T inhibits Ca^{2+}-stimulated MgATPase in reconstituted actomyosin and isolated myofibrils and decreases actin–myosin interactions. *Journal of Molecular and Cellular Cardiology*, **25**, 53–65.

Perry, S. V. (1979). The regulation of contractile activity in muscle. *Biochemical Society Transactions*, **7**, 593–617.

Pires, E. & Perry, S. V. (1977). Purification and properties of myosin light chain kinase from skeletal muscle. *Biochemical Journal*, **167**, 137–46.

Quirk, P. G., Patchell, V. B., Gao, Y., Levine, B. A. & Perry, S. V. (1995). Sequential phosphorylation of adjacent serine residues on the N-terminal region of cardiac troponin I: structure–activity implications of ordered phosphorylation. *FEBS Letters*, **370**, 175–8.

Raggi, A., Grand, R. J. A., Moir, A. J. G. & Perry, S. V. (1989). Structure–function relationships in cardiac troponin T. *Biochimica Biophysica Acta*, **997**, 135–43.

Ray, K. P. & England, P. J. (1976). Phosphorylation of the inhibitory subunit of troponin and its effect on the calcium dependence of cardiac myofibril adenosinetriphosphatase. *FEBS Letters*, **70**, 11–16.

Rayment, I., Rypniewski, W. R., Schmidt-Base, K., Smith, R., Tomchick, D. R., Benning, M. M., Winkelman, D. A., Wesenberg, G. & Holden, H. M. (1993). Three-dimensional structure of myosin subfragment 1: a molecular motor. *Science*, **261**, 50–65.

Ribulow, H. & Barany, M. (1977). Phosphorylation of tropomyosin in live frog muscle. *Archives of Biochemistry and Biophysics*, **179**, 718–20.

Ribulow, H., Barany, K., Steinschneider, A. & Barany, M. (1977). Lack of phosphate incorporation into TN-I in live frog muscle. *Archives of Biochemistry and Biophysics*, **179**, 81–8.

Smith, R. C., Cande, W. Z., Craig, R., Tooth, P. J., Schooley, J. M. & Kendrick-Jones, J. (1983). Regulation of myosin filament assembly by light chain phosphorylation. *Philosophical Transactions of the Royal Society, London, B*, **302**, 73–82.

Solaro, J., Moir, A. J. G. & Perry, S. V. (1976). Phosphorylation of troponin I and the inotropic effect of adrenaline in the perfused rabbit heart. *Nature*, **262**, 615–17.

Sweeney, H. L. & Stull, J. T. (1990). Alteration of cross-bridge kinetics by myosin light chain phosphorylation: implications for regulation of actin myosin interaction. *Proceedings of the National Academy of Science, USA*, **87**, 414–18.

Syska, H., Perry, S. V. & Trayer, I. P. (1974). A new method of preparation of troponin I (inhibitory protein) using affinity chromatography. Evidence for three different forms of troponin I in striated muscle. *FEBS Letters*, **40**, 253–7.

Tada, M. & Kadama, M. (1989). Regulation of the Ca^{2+} pump ATPase by cAMP-dependent phosphorylation of phospholamban. *Bioessays*, **10**, 157–63.

Tohtong, R., Yamashita, H., Graham, M., Haeberle, J., Simcox, A. & Maughan, D. (1995). Impairment of muscle function caused by mutations of phosphorylation sites in the myosin regulatory light chain. *Nature*, **374**, 650–3.

Westwood, S. A., Hudlicka, O. & Perry, S. V. (1984). The effect of contractile activity on the phosphorylation of the P light chain of myosin of rabbit skeletal muscle *in situ*. *Biochemical Journal*, **218**, 841–7.

Westwood, S. A. & Perry, S. V. (1982). Two forms of the P light chain of myosin in rabbit and bovine hearts. *FEBS Letters*, **142**, 31–4.

Xie, X., Harrison, D. H., Schlichting, I., Sweet, R. M., Kalabokis, V. N., Szent-Gyorgyi, A. G. & Cohen, C. (1993). Structure of the regulatory domain of scallop myosin at 2.8 Å resolution. *Nature*, **368**, 306–12.

Zhang, R., Zhoa, J. & Potter, J. D. (1995). Phosphorylation of both serine residues in cardiac troponin I is required to decrease the calcium sensitivity of cardiac troponin C. *Journal of Biological Chemistry*, **270**, 30773–80.

Chapter 6

Boyde, Y., Buckley, V., Holt, S., Munro, E., Hunter, D. & Craig, I. (1986). Muscular dystrophy in girls with X:autosomal translocations. *Journal of Medical Genetics*, **23**, 484–90.

Bulfield, G., Siller, W. G., Wright, P. A. L. & Moore, K. J. (1984). X-chromosome-linked muscular dystrophy (mdx) in the mouse. *Proceedings of the National Academy of Science, USA*, **81**, 1189–92.

Carpenter, J. L., Hoffman, E. P., Romanul, F. C. A., Kunkel, L. M., Rosales, R. K., Ma, N. S. F., Dasbach, J. J., Rae, J. F., Moore, F. M., McAfee, M. B. & Pearce, L. K. (1989). Feline muscular dystrophy with dystrophin deficiency. *American Journal of Pathology*, **135**, 909–19.

Chelly, J., Hamard, G., Koulakoff, A., Kaplan, J.-C., Kahn, A. & Berwald-Netter, Y. (1990). Dystrophin gene transcribed from different promotors in neural and glial cells. *Nature*, **344**, 64–5.

Cooper, B. J., Winand, N. J., Stedman, H., Valentine, B. A., Hofmann, E. P., Kunkel, L. M., Scott, M. O., Fischbeck, K. H., Kornegay, J. N., Avery, J. N., Williams, J. R., Schmickel, R. D. & Silvester, J. E. (1988). The homologue of the Duchenne locus is deficient in X-linked muscular dystrophy of dogs. *Nature*, **334**, 154–6.

Cox, G. A., Cole, N. M., Matsumura, K., Phelps, S. F., Hauschka, S. D., Campbell, K., Faulkner, J. A. & Chamberlain, J. S. (1993). Overexpression of dystrophin in transgenic mice eliminates symptoms without toxicity. *Nature*, **364**, 725–9.

Cullen, M. J., Walsh, J., Nicholson, L. V. B. & Harris, J. B. (1990). Ultrastructural localisation of dystrophin in human muscle using gold immunolabelling. *Proceedings of the Royal Society, London, B*, **240**, 197–210.

Cullen, M. J., Walsh, J., Nicholson, L. V. B., Harris, J. B., Zubrzycka-Gaarn, E. E., Ray, P. N. & Worton, R. G. (1991). Immunogold labelling of dystrophin in human muscle using an antibody to the last 17 amino acids of the C-terminus. *Neuromuscular Disorders*, **1**, 113–19.

Dangain, J. & Vrbova, G. (1984). Muscular development in mdx mutant mice. *Muscle and Nerve*, **7**, 700–4.

Den Dunnen, J. T., Grootscholten, P. M., Bakker, E., Blonden, L. A. J. H., Ginjaar, H. B., Wapenaar, M. C., van Passen, H. M. B., van Broeckhoven, C., Pearson, P. L. & van Ommen, G. J. B. (1989). Topography of the Duchenne muscular dystrophy gene: FIGE and cDNA analysis of 184 cases reveals 115 deletions and 13 duplications. *American Journal of Human Genetics*, **145**, 835–47.

England, S. B., Nicholson, L. V. B., Johnson, M. A., Forrest, S. M., Love, D. R., Zubrzycka-Gaarn, E. E., Bulman, D. E., Harris, J. B. & Davies, K. E. (1990). Very mild muscular dystrophy associated with the deletion of 40% of dystrophin. *Nature*, **343**, 180–2.

Erb, W. (1891). Dystrophica muscularis progressiva: klinische und pathologisch-anatomische studien. *Deutsche Zeitschrift fur Nervenheilkunst*, **1**, 13; 173.

Ervasti, J. M. & Campbell, K. P. (1991). Membrane organisation of the dystrophin–glycoprotein complex. *Cell*, **66**, 1121–31.

Ervasti, J. M., Ohlendieck, K., Kahl, S. D., Gaver, M. G. & Campbell, K. P. (1990). Deficiency of a glycoprotein component of the dystrophin complex in dystrophic muscle. *Nature*, **345**, 315–19.

Fabbrizio, E., Pons, F., Robert, A., Hugon, G., Bonet-Kerrache, A. & Monet, D. (1994). The dystrophin superfamily: variability and complexity. *Journal of Muscle Research and Cell Motility*, **15**, 595–606.

Feener, C. A., Koenig, M. & Kunkel, L. M. (1989). Alternative splicing of human dystrophin mRNA generates isoforms at the carboxy terminus. *Nature*, **338**, 509–11.

Franco, A. & Lansman, J. B. (1990). Calcium entry through stretch inactivated ion channels in mdx myotubes. *Nature*, **344**, 670–3.

Hoffman, E. P., Brown, R. H. & Kunkel, L. M. (1987). Dystrophin: the protein product of the Duchenne muscular dystrophy locus. *Cell*, **51**, 919–28.

Hutter, O. F., Burton, F. L. & Bovells, D. L. (1991). Mechanical properties of normal and mdx mouse sarcolemma: bearing on function of dystrophin. *Journal of Muscle Research and Cell Motility*, **12**, 585–9.

Khurana, T. S., Hoffman, E. P. & Kunkel, L. M. (1990). Identification of a chromosome 6 encoded dystrophin related protein. *Journal of Biological Chemistry*, **265**, 16717–20.

Koenig, M. & Kunkel, L. M. (1990). Detailed analysis of the repeat domain of dystrophin reveals four potential hinge segments that may confer flexibility. *Journal of Biological Chemistry*, **265**, 4560–6.

Koenig, M., Monaco, A. P. & Kunkel, L. M. (1988). The complete sequence of dystrophin predicts a rod-shaped cytoskeletal protein. *Cell*, **53**, 219–28.

Kunkel, L. M. (1989). Muscular dystrophy: a time of hope. *Proceedings of the Royal Society, London, B*, **237**, 1–9.

Levine, B. A., Moir, A. J. G., Patchell, V. B. & Perry, S. V. (1992). Binding sites involved in the interaction of actin with the N-terminal region of dystrophin. *FEBS Letters*, **298**, 44–8.

Lidov, H. G. W., Byers, T. J., Watkins, S. C. & Kunkel, L. M. (1990). Localisation of dystrophin to postsynaptic regions of the central nervous system cortical neurons. *Nature*, **348**, 725–7.

Little, W. J. (1853). *On the nature and treatment of deformities of the human frame*: being a course of lectures delivered at the Royal Orthopaedic Hospital in 1843. With numerous notes and additions to the present time. London: Longman, Brown, Green and Longmans.

Love, R. D., Byth, B. C., Tinsley, J., Blake, D. J. & Davies, K. E. (1993). Dystrophin and dystrophin-related proteins. *Neuromuscular Diseases*, **3**, 5–21.

Love, D. R., Hill, D. F., Dickson, G., Spurr, N. K., Byth, B. C., Marden, R. F., Walsh, F. S. & Davies, K. E. (1989). Autosomal transcript in skeletal muscle with homology to dystrophin. *Nature*, **339**, 55–8.

Matsumura, K., Tome, F. M. S., Collins, H., Azib, K., Chauoch, M., Kaplan, J. C., Fardeau, M. & Campbell, K. P. (1992). Deficiency of the 50 K dystrophin-associated glycoprotein in severe childhood autosome recessive muscular dystrophy. *Nature*, **359**, 320–2.

Menke, A. & Jockusch, H. (1991). Decreased osmotic stability of dystrophin-less muscle cells from the mdx mouse. *Nature*, **349**, 69–71.

Mimura, N. & Asano, A. (1986). Isolation and characterisation of a conserved actin-binding domain from rat hepatic actinogelin, rat skeletal muscle and chicken gizzard α-actinins. *Journal of Biological Chemistry*, **261**, 10680–7.

Monaco, A. P., Bertelson, C. J., Leicht-Galliti, S., Moser, H. & Kunkel, L. M. (1988). An explanation for the phenotypic differences between partial deletions of the DMD locus. *Genomics*, **2**, 90–5.

Ohlendieck, K. & Campbell, A. P. (1991). Dystrophin constitutes 5% of membrane cytoskeleton in skeletal muscle. *FEBS Letters*, **283**, 230–4.

Partridge, T. A., Morgan, J. E., Coulton, G. R., Hoffman, E. P. & Kunkel, L. M. (1989). Conversion of mdx myofibres from dystrophin-negative to dystrophin-positive by injection of normal myoblasts. *Nature*, **337**, 176–9.

Perry, S. V. (1992). Current status of research on the Xp21 myopathies. *Journal of Muscle Research and Cell Motility*, **13**, 377–80.

Pons, F., Augier, N., Heilig, R., Leger, J., Mornet, D. & Leger, J. J. (1990). Isolated dystrophin molecules as seen by electron microscopy. *Proceedings of the National Academy of Sciences, USA*, **89**, 2581–4.

Sato, O., Nonomura, Y., Kimura, S. & Maruyama, K. (1992). Molecular shape of dystrophin. *Journal of Biochemistry, Tokyo*, **112**, 631–6.

Sibley, J. A. & Lehninger, A. L. (1949). Aldolase in the serum and tissues of tumour-bearing animals. *Journal of the National Cancer Institute*, **9**, 303–9.

Sicinski, P., Geng, Y., Ryder-Cook, A. S., Barnard, E. A., Darlison, M. G. & Barnard, P. J. (1990). The molecular basis of muscular dystrophy in the mdx mouse: a point mutation. *Science*, **244**, 1578–80.

Suzuki, A., Yoshida, M., Yamamoto, H. & Osowa, E. (1993). Glycoprotein-binding site of dystrophin is confined to the cysteine-rich domain and the first half of the carboxy terminal domain. *FEBS Letters*, **308**, 154–60.

Walsh, F. S., Pizzey, J. A. & Dickson, G. (1989). Tissue specific isoforms of dystrophin. *Trends in Neurological Sciences*, **7**, 235–8.

Watt, D. J., Morgan, J. E. & Partridge, T. E. (1984). Use of mononuclear precursor cells to insert allogenic genes into growing muscle cells. *Muscle and Nerve*, **7**, 741–50.

Wolf, J. A., Malone, R. W., Williams, P., Chong, W., Ascadi, G., Jani, A. & Felgner, P. L. (1990). Direct gene transfer into mouse muscle *in vivo*. *Science*, **24**, 1465–8.

Worton, R. G. & Burgess, A. H. M. (1988). Molecular genetics of Duchenne and Becker muscular dystrophy. *International Review of Neurobiology*, **29**, 1–76.

Index